Cyberspace

Edited by Evon Abu-Taieh, Abdelkrim El Mouatasim and Issam H. Al Hadid

Published in London, United Kingdom

IntechOpen

Supporting open minds since 2005

Cyberspace
http://dx.doi.org/10.5772/intechopen.78887
Edited by Evon Abu-Taieh, Abdelkrim El Mouatasim and Issam H. Al Hadid

Contributors
Fabiano Belém, Jussara Almeida, Marcos Gonçalves, Chengbin Wang, Xiaogang Ma, Antonis Danelakis, Konstantinos George Thanos, Stelios C. A. Thomopoulos, Dimitrios Kyriazanos, Andrianna Polydouri, Kassu Jilcha, Evon M.O. Abu-Taieh, Issam H. Al Hadid, Ali Zolait, Carlos Pedro Gonçalves, Slavomír Gálik, Sabína Gáliková Tolnaiová, Hanaa Abdallah, Abeer Algarni, K Nalini, Jabasheela L

First published in London, United Kingdom, 2020 by IntechOpen
IntechOpen is the global imprint of INTECHOPEN LIMITED, registered in England and Wales, registration number: 11086078, 7th floor, 10 Lower Thames Street, London, EC3R 6AF, United Kingdom
Printed in Croatia

British Library Cataloguing-in-Publication Data
A catalogue record for this book is available from the British Library

Additional hard and PDF copies can be obtained from orders@intechopen.com

Cyberspace
Edited by Evon Abu-Taieh, Abdelkrim El Mouatasim and Issam H. Al Hadid
p. cm.
Print ISBN 978-1-78985-857-0
Online ISBN 978-1-78985-858-7
eBook (PDF) ISBN 978-1-78985-721-4

For EU product safety concerns:
IN TECH d.o.o., Prolaz Marije Krucifikse Kozulić 3, 51000 Rijeka, Croatia, info@intechopen.com or visit our website at intechopen.com.

Meet the editors

Evon Abu-Taieh, PhD, Associate Professor in The University of Jordan. She has authored or edited 6 scholar books and contributed to more than 8 scholar books. She has more than 55 published research studies. She is currently a visiting professor at Princess Noura Bint Abdulrahman University in Saudi Arabia. She served as Acting Dean at the University of Jordan-Aqaba for 3 years and Chair of both the CIS and BIT Departments for 2 years. She has more than 31 years of experience in higher education, computers, aviation, transport, AI, ciphering, routing algorithms, compression algorithms, multimedia, and simulation. She has served in many conferences as reviewer and on 4 journal editorial review boards. She was Editor-in-Chief of the International Journal of Aviation Technology, Engineering and Management and has been a guest editor for the Journal of Information Technology Research.

Abdelkrim El Mouatasim, born in 1973, received a Ph.D. degree in Applied Mathematics in 2007 from Mohammadia Engineering School – Mohamed 5 University in Rabat, Morocco. Currently he is an Associate Professor of Artificial Intelligence in the Polydisciplinary Faculty of Ouarzazate, Ibn Zohr University, Morocco. His research interests include: AI, optimization, mathematical modeling, and text mining.

Issam Hamad Alhadid is an Assistant Professor at the University of Jordan. In 2010 he completed his Ph.D. degree at the University of Banking and Financial Sciences in Jordan. Dr Alhadid acted as the Director of the Training and Consultation Center at The University of Jordan, Aqaba Branch. Dr Alhadid has more than 40 published research publications on Service Oriented Architecture (SOA), cloud computing, service composition AI, knowledge base systems, compression techniques, and information retrieval. Also, Dr Alhadid is a Microsoft Certified Business Management Solution Specialist and quality assurance officer, holding an EFQM certificate.

Contents

Preface

Parallel to the physical space in our world, there exists cyberspace. In the physical space there are interactions between humans and nature that produce products and services. On the other hand, in cyberspace there are interactions between humans and computers that also produce products and services. Yet, the products and services in cyberspace don't materialize—they are electronic, they are millions of bits and bytes that are being transferred over cyberspace infrastructure.

In cyberspace, just like in physical space, essential elements of interactions such as sellers, buyers, consumers, employees, students, universities, shops, and organizations do exist. But since the realm is cyberonic, a need for the imagination is essential. The idea of cyberspace is relatively new to this world, hence the need for the imagination.

This book is composed of 10 chapters that are presented in three sections. The first section, *Internet & Communications,* discusses the topic in four chapters: 5G Road Map to Communication Revolution, Cyberspace as New Existential Dimension of Human, Research and Design Methodology, and Cyberspace as New Life. The second section, *New Technologies,* is composed of three chapters: Artificial Intelligence, Natural Language Processing (NLP), and Image Processing. The third section, *Data Mining in Cyberspace*, is composed of three chapters: Text Mining that Facilitates Domain Knowledge Discovery, Tagging and Tag Recommendation, and Classifications of Posts on Social Media.

Evon Abu-Taieh and Issam H. Al Hadid
The University of Jordan,
Aqaba, Jordan

Abdelkarim El Mouatasim
Ibn Zohr University,
Ouarzazate, Morocco

Section 1

Internet and Communications

Chapter 1

5G Road Map to Communication Revolution

Evon Abu-Taieh, Issam H. Al Hadid and Ali Zolait

Abstract

The goal of this chapter is to give researchers, practitioners, and students a pedestal to get a comprehensive look at the new technology of communication named 5G. The chapter will present an introduction that shows the importance of 5G to the different uses of the Internet. Then, the chapter will present two essential aspects: (1) 5G research in academia and real world and (2) timeline of Gs. Then, the chapter will discuss three aspects of 5G which are, namely, (1) Regulations, (2) security, and (3) the 5 enabling Technologies. Then, the chapter will discuss the real-life case of South Korea mobile carrier.

Keywords: 5G, millimeter waves, small cells, massive multi-input multi-output (MIMO), beamforming, full duplex

1. Introduction

The Internet is nonpareil like running water and electricity; it is a basic need in today's world. The influence of Internet on real life reached: Economics, politics, and social. Hence, it created an alternate reality for all factors in life: e-business, e-politics, social networks, e-learning, e-culture, security, Data Science, and Big Data.

In economics, the Internet created new type of trade, i.e., e-business and new products. Some of the e-business models ranged from e-payment, new tax rules, new currency, and borderless trading. New products are different products other than the physical and non-physical products. Non-physical products include movies, music, games, and computer programs. Internet created new trades that did not exist before ranging from delivery services of physical products like delivering food, cloths, etc. to delivering non-physical products like services. According to some reports, "online retail sales worldwide will exceed US$3.4 Trillion" [1] while costs of the first year of online retailing ranges from $644 to $31,000 according to eCorner [1]. Others [2] stated that online spending is expected to reach 8.8% of total retail price increasing from 7.4% in 2016. Also, Gorlamandala [2] stated that UK has the highest retail e-commerce sales with 15.6% followed by China with 11.5%. Cross-border is an issue reported by Gorlamandala [2] listing China with $245 million and Australia, Indonesia, Singapore, France, and Mexico as countries where e-commerce sale is in noticeable numbers. Another report by Deloitte [3] stated that Amazon.com ranked fourth as top 10 retailers in 2017 with $118.573 million retail revenues with a growth of 25.3 and 36.8% retail revenue from foreign operations. Amazon in the report [3] jumped two places to number 4. The report stated "Amazon is a consistent performer in the Fastest 50, having featured in the Fastest

50 since FY2004" [3] and stated "The increase in unit sales was mainly a result of Amazon's efforts to reduce prices for customers, shipping offers, increased in-stock inventory availability, and more product variety" [3, 4]. Even Walmart is competing with Amazon over online retailing. Walmart credited the increase in sales to integrate online system with traditional sales stating "Walmart has credited its efforts to integrate its store and digital businesses so that they feed off each other." [4]. Walmart online sales have risen to 63% [4]. Furthermore, according to Wahba [5], Walmart has acquired Flipkart.com with $16 billion deal, also teaming up with Google and Microsoft and Rakuten, JD.com.

Social networks include Facebook (2.3 billion active users [6]), Instagram (1 billion [7]), VKontakte, QZone (572 million [6], Odnoklassniki, twitter (330 million [6]), Snapchat (294 million [6]), Reddit (330 million [6], Linkedin (310 million [6]), and YouTube (2 Billion [6]). Communication software include Google Duo, FaceTime, Skype WhatsApp (1600 million [6]), Facebook Messenger (1.3 billion active users [6]), Viber (260 million [6]), WeChat (1112 million [6]), and QQ (823 million [6]). All allowed connectivity and trading views in cheaper and more accessible manner. Furthermore, most of these applications are for free and allow communication for free. Still, they compensate their cost with advertising using some sort of profiling to deliver advertisement message.

Politics was affected by the cyberspace. Best example is the effect of Facebook on politics. In fact in a study by Levy [8], the researcher stated "In the measured period, stories related to politics accounted for 36% of interactions among the top 100 Facebook stories" [9], while topics like soft/general interest (17%), death (11%), science (10%), hard/general (8%), and economics (6%) followed. In fact, some revolutions were based on Facebook interaction like 2011-Jan revolution of Egypt. Another example is the American presidency race in 2016 and the Russian INTERFERENCE issue.

All the previous indicates that Internet is becoming an essential part of life and the number of Internet users is increasing rapidly. In addition, according to GSMA [10], there are 5,177,676,750 unique mobile subscribers in the world with 9,418,683,350 mobile connections. The backbone of the Internet is communication. The communication technology must meet the demand for communication. End users need more communication speed and reliability. Hence, there is a need to develop communication technology. As such, the development of 5G mobile communication technology is a promising one.

5G is the fifth generation of mobile technology that promises increased speed and lower latency, higher capacity, and higher reliability capacity [11]. 5G will be reflected in a number of today's technologies such as smart cities, connected infrastructure, wearable computers, autonomous driving, seamless virtual and augmented reality, artificial intelligence, remote robots, drones, and Internet of Machines and Things (IoMT) [12, 13]. Yet to implement 5G technologies: there are three challenges that the implementation faces: Spectrum, infrastructure, and regulations. In the next two sections, the chapter will present two essential aspects: (1) 5G research in academia and real world and (2) timeline of Gs.

2. 5G research in academia and need of real world

The topic of 5G was discussed in (14,271) research paper in conferences indexed in ACM and IEEE. Furthermore, (4542) research paper was published in journals indexed in ACM and IEEE in the past 2 years discussed topics pertaining to 5G as shown in **Table 1**. Almost 27% of the publication was published in 2019. Furthermore, one can notice that there is a race in publication pertaining to 5G from

	ACM indexed research		IEEE indexed research	
Research in:	**2019**	**2018–2019**	**2019**	**2018–2019**
Proceedings	440	748	2985	13,523
Journals	82	144	1707	4398
Newsletters	22	47		
Magazines	7	10	293	1325
Reports		1		
Total	551	950	4,985	19,246

Table 1.
Research indexed in ACM digital library and IEEE Explore pertaining to 5G.

the huge number of publications (20,196). Also, one can notice that IEEE is superseding ACM in the publications pertaining to G5.

3. Timeline of Gs

Communication technology progressed according to generations. The first generation analog communication started in the late 1970s and had a speed of 2.4 kpbs used for cellular telephones. Total access communication system (TACS), extended total access communication system (ETACS), and nordic mobile telephone (NMT) technologies were used in 1G. The main use was wireless phone call with high rate of phone drops and unclear voice.

The second generation used global system for mobile communication (GSM), general packet radio services (GPRS), and enhanced data rates for GSM evolution (*EDGE*) technologies with speed 56–64 kbps and 170 kbps when using EDGE. The second generation used digital technology rather than analog and the main uses were basic text, simple email, and snake game. The second generation included 2G, 2.5G, and 2.75G.

The third generation had four flavors: 3G, 3.5G, 3.75G, and 3.9G LTE. The speed reached 384 kbps and allowed Internet on the telephone and stream videos. 3G used universal mobile telecommunications system (*UMTS*) based on the GSM standard. While 3.5G used high speed downlink packet access (*HSDPA*) and high-speed uplink packet Access (*HSUPA*), followed by 3.75G which used high speed packet access (HSPA), an amalgamation of *HSDPA and HSUPA. 3.9G used* long-term evolution (LTE) standard.

The fourth generation 4G and 4.5G LTE reached 1 Gbps and 100 Mbps using multiple-input, multiple-output orthogonal frequency-division multiplexing (*MIMO-OFDM*). The result of such generation is HD steaming and video chats response with 0.04 ms with speed 300 MHz–3 GHz.

The fifth generation is promising millions of simultaneous connections, nearly 0 response time, massive MIMO, three times faster than 4G, and 0.001 ms response time. In short, 5G promises to be 1000 times faster than 4G. The applications of 5G are IoT, smart cities, games, autonomous cars, remote robots, drones, healthcare, and global positioning systems (GPS). Quantum cryptography for 5G security [14] is required to answer for breach of privacy in IoT. IoT, a term coined by Kevin Ashton rather than the well-known terms "embedded Internet" or "pervasive computing," will be more affected by 5G technology. Examples of objects that fall within the scope of IoT include connected security systems, thermostats, cars, electronic appliances (microwaves, fridges, washing machines, dryers, and coffee makers), lights in household and commercial environments, alarm clocks, speaker systems, and vending machines. In the next sections, the chapter will discuss the three aspects of 5G: (1) Regulations, (2) security, and technology.

4. Frequency regulations

Regulation development is required for 5G to operate. Many countries have regulations and standards for the frequency use. Hence, for 5G frequency usage, a country must develop its own regulations and standards. In the USA, according to WIA [15], only 28 states passed legislations for small cell, 3 states introduced, and the rest enacted. Laws and regulations regarding the use of frequencies need time. Hence, many countries were caught unprepared for such shift. On the other hand, countries like South Korea (2019), China, and India (2018) were already deploying the technology.

5. Security: the Prague proposal

Security is a major issue in 5G technology. The Prague proposal is none binding agreement among 32 countries from Europe, North America, and Asia-Pacific that agree on a set of security guidelines in 5G network. The countries like South Korea, Japan, Australia, New Zealand, the US, Israel, and the UK stated that security of 5G networks is "crucial for national security, economic security, and other national interests and global stability" and stresses the importance of the development of "adequate national strategies, sound policies, a comprehensive legal framework and dedicated personnel, who is trained and educated appropriately" [16].

6. The 5G enabling technologies

There are two realms that enable 5G: the physical realm and technology realm. The first realm is the physical realm as shown in **Figure 2**. The physical realm vision is to increase data traffic measured as bits per second per squared kilometer, also called *Capacity*. Capacity is calculated by Eq. (1). Capacity is the multiplication of cell density, spectral efficiency, and available spectrum.

$$\text{Capacity}\left(\text{bit/s/km}^2\right) = \text{Cell Density}\left(\text{cells/km}^2\right) * \text{Spectral efficiency} \\ (\text{Bit/s/Hz/Cell}) * \text{Available Spectrum (Hz)} \quad (1)$$

Nokia suggested to increase each element by 10×, while *South Korea Telecom* suggested to increase the first element by 56×, the second element by 6×, and the third element by 3×. In all cases, the first element *cell density* can be increased by adding more access points per km^2. The *spectral efficiency* can be increased by (1) increasing the number of antennas and (2) directing signals toward users. The third, *available spectrum*, requires the use of spectrums of 30–300 GHz, and hence, new hardware is designed and developed to handle such frequency.

The second realm is made of technologies: *Millimeter waves*, *small cells*, *massive multi-input multi-output (MIMO)*, *beamforming*, and *full duplex*.

To understand millimeter waves, one must go back in history to 1860s and 1870s when a Scottish scientist named James Clerk Maxwell developed a scientific theory that explained electromagnetic waves. The theory of Maxwell stated that electrical field and magnetic field can be coupled together to form electromagnetic waves.

"Heinrich Hertz, a German physicist, applied Maxwell's theories to the production and reception of radio waves. The unit of frequency of a radio wave—one cycle per second—is named the hertz, in honor of Heinrich Hertz" [17].

Millimeter wave is an enabling technology for 5G and refers to the use of super high frequency spectrum of 3.4 GHz. Such frequency may enable 5G technology to carry more amount of data, yet the distance is shorter, hence the need for more antennas per cell per km^2. The frequency is divided into three levels: very high, ultra high, and super high. The very high frequency ranges from 30 to 300 MHz and is mainly used in FM radio. The ultra-high frequency ranges from 300 MHz to 3 GHz and is used by TV and Wi-Fi, 2G, 3G, and 4G. The super high frequency ranges from 3 to 30 GHz [18] and is reserved to satellite broadcasting. Millimeter waves are called so because their length is 1–10 mm compared to tens of centimeters used in 4G technology [18]. mmWaves are attenuated by buildings, rain, and plants.

Because of the nature of the previously explained millimeter waves, small cells are needed. "**Small cells** are portable miniature base stations that require minimal power to operate and can be placed every 250 m or so throughout cities" [18] shown in **Figure 1**. Again, due to nature of small cells and all the interference that will be produced another technology is introduced named Beamforming. Small cells consist of small radio equipment and low-powered antennas about the size of a pizza box or backpack that can be placed on structures such as streetlights and the sides of buildings or poles. Small cells are divided into three major categories based on the coverage area, power consumption, the number of users, backhaul, application,

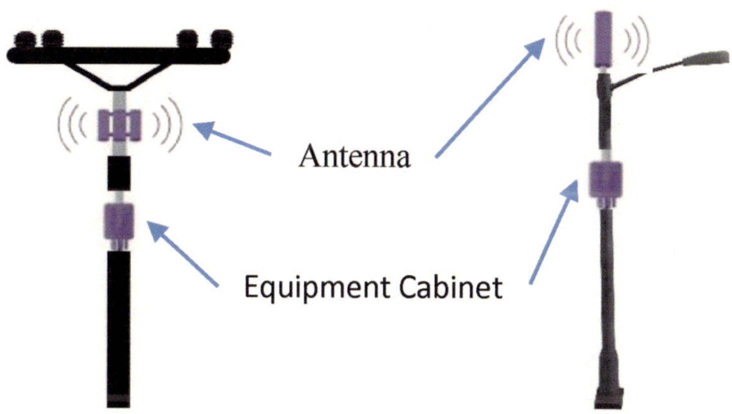

Figure 1.
Small cells mounted on power pole and streetlights.

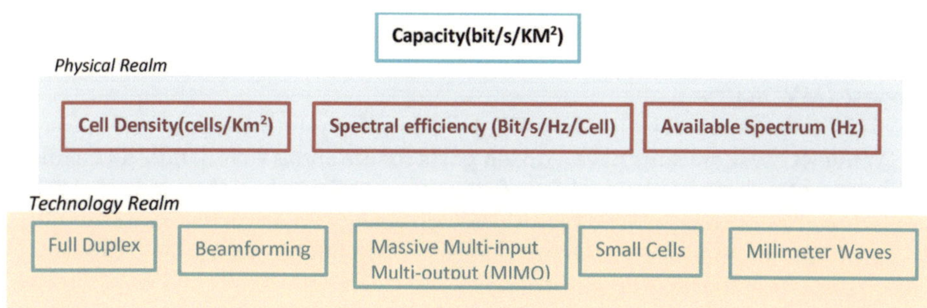

Figure 2.
Realms of 5G.

and cost: Femtocells, picocells, and microcells [19]. The coverage area of femtocell is 10–50 m, while picocell covers 100–250 m and microcell covers 500 m–2.5 km. The power consumption of femtocell is 100 mW, while picocell consumes 250 mW and microcell consumes 2–5 W. The number of users for femtocells ranges from 8 to 16 users, picocells 32 to 64 users, and microcells up to 200 simultaneous users. The backhaul of femtocells and picocells is made of fiber connection, while for microcells, fiber connection and microwave links. Femtocells and picocells are for indoor usage, while microcells are for outdoor usage. Regarding cost, both femtocells and picocells have low cost in comparison with microcells which have medium cost.

"*Beamforming* is a traffic-signaling system for cellular base stations that identifies the most efficient data-delivery route to a particular user, and it reduces interference for nearby users in the process" [18]. The major goal of beamforming is to steer a signal from communication towers and small cells to the telephone while avoiding the obstacles like building and trees, hence reducing the line drops or disconnection. "Beamforming is typically accompanied with beam steering/beam tracking. With beam steering, a transmission is dynamically adapted (i.e., steered) both vertically and horizontally by utilizing a steerable two-dimensional antenna array. By beam steering, a highly focused beam, a stronger radio signal with higher data throughput is delivered over a greater distance using less energy. The result is spectral efficiency enhancement, capacity gain, cell edge throughput gain, and mean user throughput gain" [20]. There are three types of beamforming: analog radio frequency (RF) beamformer, baseband digital beamformer, and hybrid beamforming methods; the latter is most used in 5G according to Ahmed et al. [21]. One the other hand, the traditional baseband digital beamforming (DB) requires one distinct radio frequency (RF) chain per antenna. While baseband digital beamformer has many drawbacks like the high-power consumption and high cost of mixed-signal and RF chains according to Ahmed et al. [21, 22]. The researchers of Ahmed et al. [21] conducted a comparison between digital and analog beamforming according to the following: degree of freedom, implementation, complexity, power consumption, cost, inter-user interface, and data streams. The researchers found that digital beamforming has high degree of freedom, complexity, power consumption, cost, and inter-user interface, while analog beamforming was low in the same criteria. In implementation criterion, digital beamforming used ADC/DAC while analog beamforming used phase shifters. And the data stream digital beamforming is multiple, while the analog beamforming is single. The same source lists four advantages of hybrid beamforming: (1) enabler of mmWave massive MIMO, (2) less cost for hardware and (3) operation, and (4) energy efficiency. Ali et al. [23] added two more advantages: (5) Improved spectral efficiency and (6) increased system security. Ali et al. [23] listed the following algorithms used in beamforming: least-mean-square (LMS) [24]; recursive-least-square (RLS); sample matrix inversion (SMI) [24]; conjugate gradient algorithm (CGA); constant modulus algorithm (CMA); least square constant modulus algorithm (LS-CMA); linearly constrained minimum variance (LCMV); and minimum variance distortion less response (MVDR).

MIMO is the technology used by 4G and stands for multiple-input multiple-output. While 4G base stations have a dozen ports for antennas that handle all cellular traffic: eight for transmitters and four for receivers, 5G can handle hundreds [18] and is duped as *massive MIMO*. To achieve such goal, 5G must install more antennas which will produce more interference, hence the need to beamforming. Massive MIMO systems will utilize beamforming.

Full duplex is the technology that allows a transceiver to send and receive data simultaneously [18]. To achieve such goal, researchers must design hardware that will allow antennas to send and receive simultaneously. "To achieve full duplex in personal devices, researchers must design a circuit that can route incoming and outgoing signals so they don't collide while an antenna is transmitting and receiving data at the same time" [18]. "*One drawback to full duplex is that it also creates more signal interference, through a pesky echo. When a transmitter emits a signal, that signal is much closer to the device's antenna and therefore more powerful than any signal it receives. Expecting an antenna to both speak and listen at the same time is possible only with special echo-canceling technology*" [24].

7. Current situation—South Korea

Currently, 5G is facing many challenges to be implemented; the following is the case of South Korean mobile carrier. On the 20th of March 2019, South Korean mobile carrier (SK Telecom) announced using quantum cryptograph technology for the security of 5G network. SK applied quantum number generator (QRNG) technology of ID QUANTIQUE (IDQ) for 5G subscribers to prevent hacking and eaves dropping. SK invested $65 million into IDQ and plans to expand the use QNRG. Furthermore, SK wants to apply quantum key distribution (QKD) technology in April/2019 [14, 25].

8. Conclusion

There are many publications and published research (20,196) that pertain to 5G technology. This chapter gives researchers, practitioners, and students a pedestal to get a comprehensive look at the new technology of communication named 5G. The chapter first gives an introduction about the increasing need for 5G technology. Then, it shows the amount of research conducted and indexed in ACM and IEEE. Next, the chapter shows the development of telecommunication technology from first to fourth generation. The chapter discusses three important aspects of 5G: Regulations, security, and the five enabling technologies. The five enabling technologies included two realms: physical realm and technology realm. The physical realm included discussion of capacity, cell density, spectral efficiency, and available spectrum. On the other hand, the second realm is made of technologies: *Millimeter waves, small cells, massive multi-input multi-output (MIMO), beamforming*, and *full duplex*. The seventh section presented current situation—South Korea mobile carrier.

Author details

Evon Abu-Taieh[1*], Issam H. Al Hadid[1] and Ali Zolait[2]

1 Department of Information Systems, College of Information Technology, The University of Jordan, Jordan

2 Department of Information Systems, College of Information Technology, University of Bahrain, Kingdom of Bahrain

*Address all correspondence to: abutaieh@gmail.com

References

[1] eCorner. Costs & issues starting an eCommerce business or online store. eCorner Pty Ltd [Online]. 11 April 2020. Available from: https://www.ecorner.com.au/FAQs-Questions-Ideas-Centre/Costs-Issues-Starting-An-eCommerce-Online-Business

[2] Gorlamandala R. Retail Disruption in a global-digital era. TatCAPITAL Corporate Finance & Advisory [Online]. 19 November 2018. Available from: https://tat.capital/Home/Permalink?slug=Retail-Disruption-in-a-global-digital-era

[3] Deloitte. Global Powers of Retailing. Deloitte [Online]. 2019. Available from: https://www2.deloitte.com/content/dam/Deloitte/global/Documents/Consumer-Business/cons-global-powers-retailing-2019.pdf

[4] Amazon. Annual report. Amazon.com, Inc. [Online]. 2018. Available: https://sec.report/Document/0001018724-18-000005/amzn-20171231x10k.htm

[5] Wahba P. Walmart's U.S. Online Sales Rise 63%. 18. Fortune [Online]. 18 May 2017. Available from: http://fortune.com/2017/05/18/walmart-online/

[6] Clement J. Most famous social network sites worldwide as of July 2019, ranked by number of active users (in millions) [Online]. 2019. Available from: https://www.statista.com/statistics/272014/global-social-networks-ranked-by-number-of-users/

[7] Clement J. Number of Monthly Active Instagram Users from January 2013 to June 2018 (in Millions). Statista; 2019

[8] Levy P. Collective Intelligence: Mankind's Emerging World in Cyberspace. Cambridge, MA, USA: Perseus Books; 1997

[9] Clement J. Distribution of top 100 stories on Facebook worldwide as of March 2019, by genre, Statista [Online]. 2019. Available from: https://www.statista.com/statistics/1001970/top-100-stories-facebook-genre-worldwide/

[10] GSMA. DATA DASHBOARD, GSMA Intelligence. 2020 [Online]. Available from: https://www.gsmaintelligence.com/markets/1826/dashboard/

[11] Czentye J, Dóka J, Nagy Á, Toka L, Sonkoly B, Szabó R. Controlling drones from 5G networks. In: Proceedings of the ACM SIGCOMM 2018 Conference on Posters and Demos (SIGCOMM '18). New York, NY, USA; 2018

[12] Barak S. 5G is So Near-Future: A Look Ahead to 6G and 7G. ICONS of Infrastructure [Online]. 2018. Available from: https://iconsofinfrastructure.com/5g-is-so-near-future-a-look-ahead-to-6g-and-7g/ [Accessed: 10 January 2020]

[13] Katz M, Matinmikko-Blue M, Latva-Aho M. 6Genesis Flagship Program: Building the Bridges Towards 6G-Enabled Wireless Smart Society and Ecosystem. In: 2018 IEEE 10th Latin-American Conference on Communications (LATINCOM). Guadalajara, Mexico; 2018

[14] Chau F. SK Telecom using quantum cryptography for 5G security. telecomasia.net [Online]. 2019. Available from: https://www.telecomasia.net/content/sk-telecom-using-quantum-cryptography-5g-security [Accessed: 01 January 2020]

[15] WIA. 28 States Passed Small Cell Legislation for 5G Deployment. Wireless Infrastructure Association [Online]. 2019. Available from: https://wia.org/smallcelllegislation/

[16] Bushell-Embling D. 32 countries agree on 5G security guidelines. telecomasia.net [Online]. 2019. Available from: https://www.telecomasia.net/content/32-countries-agree-5g-security-guidelines [Accessed: 01 January 2020]

[17] Science Mission Directorate. Anatomy of an Electromagnetic Wave. National Aeronautics and Space Administration [Online]. 2010. Available from: http://science.nasa.gov/ems/02_anatomy

[18] Nordrum A, Clark K. Everything You Need to Know About 5G. IEEE Spectrum [Online]. 2017. Available from: https://spectrum.ieee.org/video/telecom/wireless/everything-you-need-to-know-about-5g [Accessed: 25 December 2019]

[19] Rajiv. What are small cells in 5G technology. RF Page [Online]. 2018. Available from: https://www.rfpage.com/what-are-small-cells-in-5g-technology/

[20] Ericsson. Beamforming, from cell-centric to user-centric. Ericsson [Online]. 2019. Available from: https://www.ericsson.com/en/networks/trending/hot-topics/5g-radio-access/beamforming [Accessed: 25 December 2019]

[21] Ahmed I, Khammari H, Shahid A, Musa A, Kim KS, Poorter ED, et al. A survey on hybrid beamforming techniques in 5G: Architecture and system model perspectives. IEEE Communication Surveys and Tutorials. 2018;**20**(4):3060-3097

[22] Ahmed I, Khammari H, Shahid A. Resource allocation for transmit hybrid beamforming in decoupled millimeter wave multiuser-MIMO downlink. IEEE Access. 2017;**5**:170-182

[23] Ali E, Ismail M, Nordin R, Abdulah N. Beamforming techniques for massive MIMO systems in 5G: Overview, classification, and trends for future research. Frontiers of Information Technology & Electronic Engineering. 2017;**18**(6):753-772

[24] Clement J. Worldwide mobile app revenues in 2014 to 2023 (in billion U.S. dollars). Statista [Online]. 2019. Available from: https://www.statista.com/statistics/269025/worldwide-mobile-app-revenue-forecast/

[25] SK Telecom Continues to Protect its 5G Network with Quantum Cryptography Technologies, IDQ [Online]. 2019. Available from: https://www.idquantique.com/sk-telecom-continues-to-protect-its-5g-network-with-quantum-cryptography-technologies/ [Accessed: 01 January 2020]

Chapter 2

Cyberspace as a New Existential Dimension of Man

Slavomír Gálik and Sabína Gáliková Tolnaiová

Abstract

Since the second half of the twentieth century, especially from the 1990s to the present, we have seen significant sociocultural changes that have mostly been influenced by information technology. In the area of information technology, it is mainly the Internet that is the essential part of all modern communication technologies such as smartphones, iPads, and so on. The Internet is a new communication space, also called cyberspace, in which we not only communicate but also work, learn, buy, have fun, and so on. It does not seem to be a mere "tool" of our new way of communication, but a dimension that becomes part of our existence. We then have to ask how our existence is changing under the influence of new technologies. How do we change the value system in cyberspace communication? What are the possibilities and risks of communication in cyberspace? These are just some of the issues that arise in connection with communication in cyberspace to which we will seek answers. In the chapter we use the phenomenological and hermeneutic method. Through the phenomenological method, we examine the basic structure of cyberspace (Clark, Ropolyi) and, using a hermeneutic method, examine the differences between communication in cyberspace and old media (Lohisse, Postman, Bystřický).

Keywords: cyberspace, communication, existence, information, time, space, thinking

1. Introduction

In the beginning of the chapter, we want to clarify the concept and basic structure of cyberspace. The term cyberspace was used for the first time in 1984 by a sci-fi writer, Ford Gibson, in his novel *Neuromancer*. The etymology of this word reveals that it describes a cybernetic space that is not identical with three-dimensional physical space; it is a place that merely simulates the real one. Simulations may be visual or acoustic but also more sophisticated, for example tactile, when a special pair of sensory gloves is used. Cyberspace is constructed using communication technologies, particularly the Internet. Sometimes cyberspace and the Internet are understood as identical places. However, we think that the term cyberspace embraces more than just the Internet. We agree with D. Clark [1], who notices that we need to use the word cyberspace even in connection with discovery of telegraph. Two people in two different points on the globe enter the acoustic communication space that does not share three dimensions. However, modern media offer a superior kind of cyberspace—the Internet—that we can use to communicate not only acoustically but also visually, through services such as skype, for example. We also need to distinguish cyberspace from other kinds of space, for example, social, economical, or mental space.

We should also distinguish cyberspace form virtual reality. The term "virtual reality" is used to describe something artificial, constructed, or less real. In contrast, cyberspace does not necessarily mean something unreal. When we make a phone call or use skype to communicate with somebody, we do not take it as something virtual, despite the fact that this communication takes place in cyberspace. Another difference is also in the sense of emphasis. Virtual reality emphasizes reality, while in the case of cyberspace, it is the actual place that we emphasize. Though the term reality incorporates also space, it spreads even further and gets closer to the philosophical term of being—existence or possible existence. Based on understanding such elementary terms as reality and space, the terms of virtual reality and cyberspace are derived from them. The meaning of virtual reality will be defined by the relation between reality and virtual reality and that of cyberspace by the relation space-cyberspace.

D. Holmes [2], in reference to Ostwald, states that cyberspace means communication space of a number of people: "individuals do not exist in cyberspace, but in virtual reality." In this case the term of cyberspace is limited to exist only for social communication. But why could not we call this space cyberspace, when all kinds of communication, individual or collective, happen in the same technological space? Why could not we call this space cyberspace, when all kinds of communication, individual or collective, happen in the same technological space? We believe we should think of cyberspace as of the traditionally understood physical space and not solely as something that we derive from social relationships; we should see it in a more contextual sense—in the sense of relations between objects. We borrow this term and use it to express the mental space in which we think and construct and then transfer our constructions into the technological world that could be properly seen as cyberspace.

In order to understand cyberspace better, it might be useful to learn more about its internal structure, or more precisely its hierarchy-based levels. According to Clark [1], there are four levels in cyberspace: physical level, logical level, information level, and human level.

The physical level of cyberspace is composed of physical devices that are interconnected. These are computers, servers, sensors, transmitters, the Internet, and communication channels. Communication "flows" between these technical devices through cables, optical fibers or electromagnetic waves. This physical level is the easiest to touch physically, especially devices, as they are easy to locate.

Cyberspace, according to Clark, is built from various components starting with the lowest level and ending with the highest one. The lowest level is represented by a program that performs basic operations, data transfer, and formatting. These services serve applications such as database or web. For example, by combining database and web, we get creative and active web content. On the top position of the web, we can find services such as Facebook, which is yet another platform for further applications. The essence of cyberspace lies in continual and rapid increase of possibilities and services that are based on creation and combination of new logical constructions. Cyberspace, as a logical level, then means a series of platforms for new creations and constructions that consequently become innovations. Cyberspace is very flexible and recursive, building platforms on further and further platforms.

Clark believes that creation, obtaining, and transfer of information are the essential functions of cyberspace. Information here takes various forms, for example, music, video, or websites. Information about information is generated here (metadata), and information that informs about other information is produced, for example, by Google. The character of information in cyberspace changed with computers being connected to other computers; they started processing the structure of data. Data is saved not only statically on hard drives or USB memory sticks, but more and more it is created dynamically on networks, where physical localization loses its importance.

Clark sees the highest level of cyberspace in people, who are not only passive users but also contributors to the content that it offers. If people contribute to Wikipedia, then Wikipedia exists. If people tweet, then also Twitter exists. Cyberspace is meant to serve people, for communication, for a content that is constantly refreshed. This is the reason why people are the most critical component in cyberspace. Also Cocking [3] sees it similarly, when he claims that today people present themselves more and are more engaged in various activities with others through computer technology. We can also state that the Internet, or cyberspace, offers a great way to express oneself and communicate in a modern society [4].

Ropolyi [5] contemplates the Internet in similar intentions, though with small variations. Ropolyi understands the Internet as a complex, multilayered system in which four levels are identified: the technological, communicational, cultural, and organismic levels. Ropolyi says—with respect to the first level—that the Internet is a system of computers that are able to rapidly and securely access information inside the worldwide network. As a technological tool, the Internet is connected to other technological tools that support different human and social needs, ranging from shopping to international financial transactions. Ropolyi understands the next level of the Internet as a space for different types of communication. He believes that the Internet represents an active agent within such communication, as it only facilitates, prompts, and enables specific types and forms of communication. A range of content, including text, audio, and image, can be communicated over long distances thanks to the Internet. The third level, according to Ropolyi, is cultural, which must be understood in the widest possible sense and which contains different human ambitions, intentions, values, plans, and products. The Internet as a universal medium may grasp the same cultural values and activities as the real world. It also creates a new cultural world in which self-realization can be accomplished in many different ways that would simply be impossible in the real world. Finally, according to Ropolyi, the Internet is an independent organism that can be examined separately from the technology inside its structure. This globally distributed organism develops in the same manner as any other evolutionary system. People themselves, along with their thoughts, actions, and ambitions, are a part of this organism.

The structure of cyberspace represents a hierarchy-based system of technical and semantic layers (physical, logical, information, and human) that are heavily linked to each other. The most important goods in this space are information, which is used by people, thus creating their new living space.

2. Information as the basic unit of cyberspace

Information that is stored in cyberspace can be seen as its basic building block. Information, or more precisely communication of information, builds up the very cyberspace, and without this building block, cyberspace would remain just a possible construction (*in potentia*). Cyberspace thus presents a platform for communication of information, rather than an independent entity.

What precisely is information that we communicate in cyberspace? We can see it as a correlation of two entities: physical and conceptual. By physical entity we can mean, for example, computer hardware or radio waves. Information, regardless of physical media that is used to spread it, is coded in a binary code or "binary digit" (0 and 1). New communication technologies that are based on a binary number system are therefore known as digital media. Meanings are programmed and stored in computers as data, which can in semiotic transcription represent text, sound, images, and so on. Correlation of the physical and nonphysical world is well known in linguistics and semiotics. For example, human speech is a correlate of sounds (phonemes) and meanings. Correlation of the material and nonmaterial world is well known, for example, in linguistics or semiotics. For example, human speech is a correlate of sounds (phonemes) and meanings. Human articulated sounds, if performed in the correct order, can be decoded, and their meanings can be understood. Also handwriting is a correlate of signs and meanings. If symbols are written syntactically and grammatically correctly, then they can be decoded, which means they can also be understood. The difference between human language and "talking" through cyberspace lies in the fact that material correlate of information is constructed using modern technologies. Information coding, in contrast to speech and writing, is not performed directly and immediately, but through modern technologies in the form of binary code. This "information" cannot be approached semiotically and, as such, lies outside the human natural comprehension—this information we call simply "data." Information that people work with has already been processed by computers and therefore is regarded as proper information. Ontologically, we need to distinguish data and information. Data is composed of binary values, with a given functional structure, while when processed by computer system, it is turned into information. For instance, data that represents a rose become information about the very rose when this data is transposed into the human semiotic system. The image of rose will then be matched to certain ideas, desires, and so on. Thus, information is always richer than simple data that is formed by logical and functional algorithms.

Růžička [6] and Cejpek [7] distinguish data and information similarly. They mention one more important difference between information types. Růžička [6] explains: "One can speak of data when the world is measured, weighed, counted...." For him, the structure of facts is less formal than that of data: "Facts are less formalized than data, but they can still be deprived of context. ...Data is the result of a mathematical formula. ...In my opinion, factum is a testimony, a description of the world that, in a given scope, is problem-free and undisputable." He talks about quality of information: "...neither data nor facts is identical with information. The quality of information will reveal when a dialogue comes in which the world in question is challenged, or: when facts talk, questions are needed...."

Similarly, Cejpek [7] distinguishes between information and knowledge, when he says: "...in a more detailed view of given information, we need to distinguish between information and knowledge." Or, as he continues: "Information as such does not mean recognition, but constitutes certain pre-requisite and basis." He bases his idea on Patočka, philosopher [7], who claimed that "the term of information cannot explain understanding and knowledge...."

When compared, we can place Růžička's facts to Cejpek's information and Růžička's information quality to Cejpek's knowledge **Table 1**.

Name	Information type 1 outside of semiotic system	Information type 2 isolated information	Information type 3 information in context
Cejpek	Data	Information	Knowledge
Růžička	Data	Fact	Information quality

Table 1.
Comparison between two similar semantic approaches to structure of information.

The first contact with information that comes from a data source may appear to be isolated and simple or even measurable. Both authors emphasize that we can only count data blocks, not information blocks. Essentially, even information seen for the first time is not isolated or unbreakable, as we use semiotic rules, based on a system of relations, to understand it. However, both of these authors agree that deeper assessment brings higher information quality or knowledge that affects us, since it widens and re-configures the horizon of our knowledge. Such deeper understanding then brings serious consequences not only into the way we understand given information but also knowledge or knowledge-based society.

Information is a basic, ontological unit in cyberspace and can be also seen through the lens of classical metaphysics. Similarly to Aristotle's metaphysics, also information is made of its matter (material correlate) and form (idea correlate). It has its potency as data and validity as information or information quality.

3. Cyberspace as a new existential dimension of man

If a significant part of our life, for example, our visions and ideas, is reflected in cyberspace, then we can say it becomes a new extension of our life. If we daily spend a few hours in cyberspace, then the bond with our life will be very strong. Lohisse [8], points out that media (including cyberspace, as a communication channel for modern digital media—note by authors) are not mere tools that do their job only when we use them, but they expand and their effect grows. More specifically, this influence can be seen in the adaptation of our cognitive functions and abilities (attention, memory, imagination, thinking, etc.) to cyberspace communication. And this adaptation changes our existence. Our existence extends to a new dimension that is virtual in nature. The virtual dimension, or the cyberspace in which we communicate, thus becomes a new existential dimension of man.

The very first thing that will attract our attention when we study the phenomenon of cyberspace is its character. Paradoxically, we can describe cyberspace as a non-space place, as there is no 3D physical dimension in it. Despite this feature, we still regard it as a space, even though we mean it predominantly in a visual or audiovisual sense. Thus, this new technological space lies within a human, in the very mental dimension we use for constructing vision or ideas. The difference is in the fact that human's mental space is given biologically, while cyberspace is constructed technologically.

The second thing that may attract our attention in communication in cyberspace is the speed of communication. Communication is almost instant, typically with no delay. Besides this, there are no firm physical marks that could be used for distinguishing movement, which is something we need when we want to measure time. Immersed in cyberspace, we are not able to measure time. In order to do it, we need to step outside. Events in cyberspace resemble a dream in which we cannot say time. Cyberspace and dreams both share two features—no fixed points that could be used for measuring and no perspective for the observer. When we dream, we first need to wake up, only then we can measure the time spent. With new technology, for example, Google Glass or electronic lenses, leaving cyberspace would not be so easy because Google Glass, or let alone electronic lenses, would be quite an integrated part of the human body.

The speed of communication and absence of physical space in cyberspace eliminated linear or successive time. We could also call it simultaneous time, borrowing the term from a simultaneous exhibition in which the grand master plays multiple games of chess at a time with a number of players. The idea of linear or gradual time breaks up into a pattern of present events. Something similar happens also in communication in cyberspace, for example, when we surf the Internet [9].

Time and space are two basic coordinates of our life, marginalization or omission of which can greatly affect our life. According to I. Kant, time and space represent a priory aesthetical forms of consideration, the first and fundamental processing of impressions that we get through our senses. If this is changed, then there is a great chance that our everyday real life will get changed as well. Time and space will not be as important as they use to be. For example, a medieval man saw time as a gift; it meant a chance to fight for salvation. In the modern period, time might have meant a space for self-realization. Nowadays, influenced by cyberspace, time not only becomes "just now," but it is also empty. The result of time made present is seen in the youngest generation as a lack of interest in history, but also future, as these people live their lives more and more in chatrooms, on Facebook, sharing photographs, videos, and other similar experiences. In such a social space, information about the past but also future, about plans or vision, would feel very disruptive. Rankov [10], inspired by Lévy, comments that time (with tradition and culture) spreads into hypertext, which we read not linearly, but consecutively. In other words, information that was once spread is now stored in database or in cyberspace, where it is distributed, combined, and broken into chunks. Also, time is not the same as it was in the past. Despite the fact that everything speeds us and modern society suffers from chronic lack of time, we are killing the time more and more by surfing on the Internet, useless chatting, or sending emails.

Similarly to time, also space—or more precisely our ideas of space—have changed. We take space very differently from how we understood it in the past, for example, in the Middle Ages or Modern Period time. A man in the Middle Ages could learn about distance between, for example, Rome and Paris by actually walking or riding a horse from one place to another for 3 or 4 weeks. His experience of the distance would be equal to the trouble he went through during this journey. In the Modern Period, with the discovery of America sailing all over the globe, the idea of space was changed. Though our Earth was still huge, it was not limitless as it was a sphere. In the nineteenth and especially twentieth century, with development in modern transport and information technology, the Earth became even smaller. We can travel to the most distant places within hours, and when we use telecommunication technology, we can make this journey in an instant. Telecommunication technology (auditory and visual) eliminates physical dimension in space. We take this form of online communication as an absolutely standard service and do not realize the loss of real space.

Referring to I. Kant's epistemology, with aesthetic forms of outlook, such as time and space, also our category of thinking changes. Kant distinguishes 12 categories as an a priori matrix that contributes to our thinking. In more recent philosophy, influenced especially by L. Wittgenstein and M. Heidegger, a discovery was made—our thinking, including category pattern, is firmly bound to our language. This means we think and learn in the language that we communicate in. Spoken word is understood to be a privileged medium, mother of all media. However, it is not the only medium as we also use written word, printed word, and electronic media, including the Internet—which we generally use to enter cyberspace. If we then think with media, then each kind of media must affect the form or structure of our thinking. Lohisse [8] provided convincingly evidence on how thinking (collective mentality) was influenced by four types of media through the cultural history of mankind: spoken word, written word, printed word, and electronic media. According to him, spoken word was potent to draw and unite people deeply. The era of spoken word featured cyclic time and collective consciousness. This was broken with the beginning of written word. Writing, especially phonetic one, reorganized human thinking into a linear template, which also initiated a shift to linear understanding of time. Written word became a tool to divide the society that started to see the phenomenon of power and individuality. This trend was even more evident in the era of printed word, which separated the author and established a standardized text, fostered individuality of man and subject-object view, and also triggered the mass phenomenon. Lohisse sees electronic media, but specifically the Internet, as fundamentally different, changing our imagination and the way we think and learn. The Internet uses a technological language, and we have to adapt to this language in our communication. Our language will therefore be changing into a techno-language. Besides this, speed and amount of information will be shaping our thinking toward discontinuity, simplicity, and superficiality of content. On the other hand, the Internet might give us a chance to improve our skills to quickly respond to varying content that we find in cyberspace, which is something our predecessors would probably have a problem with. It is rather difficult to map how thinking of a modern man changes, but it becomes apparent when compared with people in history. N. Postman [11] offers an impressive example of a nineteenth century dispute between Lincoln and Douglas. They both were able to maintain their debate on an exceptionally high rhetorical level for long hours and keep their audience interested. They could still continue their debate after a longer break. Postman showed the contrast with television, which through often miss-matching images deforms abstractive thinking, once highly cultivated by printed word. Pravdová [12] points out that "it is enough when images, can be distinguished, in contrast to words, which need to be understood." A similar situation happens also in the era of the Internet. The cyberspace Internet favors image thinking, unconcentrated and not too continual logically. In the context of these changes, Sartori [13] points out that man changes anthropologically and *Homo sapiens* turns into *homo videns*, which testifies to change from abstract to image thinking.

Communication in cyberspace triggers changes in understanding time, space, and structure of thinking. In order for us to communicate in cyberspace, we need new information technologies; these become an everyday part of our life. This is yet one more effect that cyberspace brings. Originally, modern information technologies were not mobile, just as the heavy computers we saw in the 1990s. With light and small notebook computers and presently also iPads and smartphones, this technology is easy to carry. They are part of our life not only at home but also in the streets, offices, and generally in any possible place that we go to. These modern devices that help us enter cyberspace are generally at hand. With Google Glasses, which do not require physical manipulation, cyberspace becomes somehow a part of the body. Google has a vision—such glasses could be transferred into electronic lenses. This would mean a very close bound between body and modern information technologies. With these communication changes, we start thinking about cyborgs, where technologies become a part of the human body. With everyday usage of smartphones or iPads and physical connection between them and the human body (they are at hand, in the pocket, etc.), we can start speaking of mental cyborgism because combining the human body and technologies happens at a mental level. However, if such technologies became a real part of the human body, it would mean real cyborgism, or direct connection of the human body and technology. We agree with R. Cenká and I. Lužák [14] that "technosphere is taking over biosphere" and that this trend will continue. This makes us wonder what will happen with human naturalness. Will we still be able to talk about the old human, or will it be a new kind of human? These questions might look like a sci-fi, but in a few years' time or decades, they could describe reality.

Another problem with identification with media is the one of cyberspace identifications with social groups or one's own avatar. It is not quite about what social group or what avatar it may be, but about the need to get somehow inside a group, identify oneself with the group, or change identity. Our identity can then be constructed in accord with our participation in various groups.

We can call the changes that we studied in this part of subchapter (changes in time, space, structures of thinking, and identification with technologies) formal, because they are results of using mental or physical connection to information technologies. Of course we could mention other formal influences, and we may, for example, study changes in the attention, memory, social contacts, and more. In the background of this approach is the idea of technological determinism, such as M. McLuhan, L. A. White, J. Lohisse, and other authors. The starting point for this approach is the idea that new communication technologies have a profound impact on human cognitive changes and consequently changes in culture and society. Along with formal influences, there are also changes based on content influence. We see content as particular communicated information that may take various forms—perhaps as symbols (images, sounds, and so on) or meanings (scientific, social, entertaining, and similar). Formal influence of cyberspace, though harder to be recognized as it is not a direct product of communication, has a stronger influence on shaping a man than communicated content, because it structurally changes his ideas and thinking. S. Gáliková Tolnaiová [15] calls the formal type of influence stronger version and the second, content type, weaker version of media influence. It is chiefly the first—formal type—that contributes to the new anthropogenesis, influencing man mentally, psychically, and also physically to certain degree.

4. Positives and negatives of formal influence of cyberspace

The usage of information technologies and especially communication in cyberspace has its positives but also negatives. Modern communication technologies, similarly to other tools, can act as a good servant but a bad master. It is very difficult today to find the borderline between these two polarities because the bound between us and them is so strong that we are more or less unable to reliably distinguish and realize how much they influence us. Middle-aged and older generation, having lived without the influence of modern media, is more likely to debate this than younger, or the so-called digital generation, as they were growing up surrounded by new technologies that became an inseparable part of their life. Therefore it is extremely important to learn to see the perspective, build up a mental a psychological protective barrier when approaching media, and distinguish positives from negatives in communication in cyberspace.

We can now speak more on positives and negatives of the four formal influenced areas (time, dimension, structure of thinking, and identification) in cyberspace:

1. Time. When we communicate in cyberspace, we experience enormous speeding up of information transfer, which nowadays reaches almost the speed of light. Then there is a huge increase of amount of information, which still grows exponentially. This means we can access almost any information quickly, but selecting and processing are more demanding and time-consuming, which lead to sketchiness. The lack of time further causes another effect—deprivation of time that should be dedicated to holiday, family, bringing up children, and so on. Paradoxically, one may be killing the time by surfing the Internet, chatting to friends, or sending emails simply to maintain the feeling of being engaged or belonging to a group. Besides this, information in cyberspace is not stored and communicated linearly, but hypertextually, in a fanlike pattern, which leads to favoritism of simultaneous time over linear time. With linear time being broken comes also lack of interest in the past, history, culture, and traditions but also indifference regarding the future. This is typical of the modern digital

youth. Bauerlein [16] carried out research at high schools and universities in the USA and found out that year after year students are less and less aware of history and civic education, and generally their knowledge in subjects that have something to do with history is less and less adequate. Volko [17] carried a knowledge research at one of the Slovak universities and acquired a similar result, which he commented: "Quality of general knowledge is, mildly speaking, inadequate. Students that will in the future work in media, struggle when asked to say for example when Slovak National Uprising started, they cannot define holocaust or think of two Slovak classical music composers."

2. Space. A positive aspect of communication in cyberspace is in its ability to defeat geographical locations. We can now communicate with someone who lives in Australia or the USA not only orally but also visually. We even can watch events happening in various places on the globe. This may bring its negative aspect—we can lose the sense of value of the real surrounding, our homeland, traditions, and culture in a given place. With communication on the Internet, importance of such a place declines, and people lose their roots. The Internet and also globalization tear the bound between geographical location and social role. With no geographical and social roots, one can easily become homeless in cyberspace.

Cyberspace of electronic media does not only consist of online telecommunication or online tele-seeing of the world; it is a world of new opportunities in virtualization of reality. Virtualization of reality may take various modes of reality or creation of brand new, fantastic worlds. Communication, or contact with virtual worlds, brings some pros and some contrast. In playing games young people can learn manual and visual skills and learn about the world but also become completely immersed and become addicted or virtualize the real world.

3. Structure of thinking. In communication within cyberspace, some structural changes in thinking and consequently in learning occur. Each media has its own semiotics, and the most fundamental media, for example, spoken word, found their new cultural epochs. Therefore, media are tools for our thinking and learning. Bystřický [18], for instance, says that "we also use different ways of thinking with increased use of technologies, not in terms of changing the actual availability of such abilities; we rather fundamentally alter strategy of their use." Thinking in cyberspace is influenced by discontinuity of images, short texts, and similar, which does not help us to train concentration and continual refinement of ideas. On the contrary, a text in a newspaper or book requires us to concentrate and pursue the logical chain of ideas that are expressed. Book and newspaper thus develop abstract and logically continual thinking, while television and especially the Internet nourish visual and discontinuous thinking. According to G. Sartori, image-based media, such as television and the Internet, alter the way we think, imagine, and learn. He is convinced that a new type of human is rising—*homo videns*—whose perception and knowledge are greatly modified by media images. In his idea, the turn from conceptual language of texts to media images also brings deprivation of abstract thinking, and, as Solík [19] adds, also emotional changes. We do not need to think; seeing a picture is enough. Sartori [13] explains: "Television brings metamorphosis that affects the very core of *Homo sapiens*. It is not a mere tool for communication, but also an anthropological instrument that constructs a new kind of human existence." *Homo sapiens* then changes to *homo videns*, which introduces decline in erudition and cultural decadence. It is similar in the case of the Internet that, unlike television, is interactive. If used by

someone culturally illiterate, only what seems to be interesting will be picked, namely, entertainment. We could see this aspect of structuring in thinking as negative. Positive aspect could be seen in rapid access to information and, under certain condition, also access to information offered by "collective intelligence," collective source of information, for example Wikipedia.

Another structural change in thinking in communication in cyberspace occurs in net-based or hypertext-based source of information. We could describe this type of communication or information as rhizomorph. Eco [20] used this term to distinguish it from the previous, treelike (*arborescent*) thinking. A picture of tree, for example, in the Middle Ages thinking (*arbor porphyriana*), represented a neat structure of hierarchy-based and logical thinking, from the essence of being, all the way down to its peripheral symptoms. However, rhizomatic thinking is non-systematic, incomplete, and netlike and has no beginning and no end. Thus the Internet, based on its own technological and netlike (rhizomatic) structure, promotes "loop connections" and consequent disintegration of the so-called linear code. U. Eco explains that rhizome excuses and supports disharmony, because rhizome creates loop-like processes. Eco even says that "To think means, in rhizome, to advance blindly and rely only on assumptions." The Internet, characteristic for its hypertext, or perhaps rhizomatic connection of information, will not support abstract, linear, and logical thinking, which may constitute a threat for modern society. Spitzer [21] states that digital natives instead of thinking in hermeneutic circle (from fragments to the whole picture and the other way round) get only superficial information surfing on the Internet: "Digital natives do not go through this hermeneutic circle: they haphazardly click here and there and never return to a good source; they look only horizontally (do not dig deeper)." Višňovský [22] notices that there is a difference between printed information and online information. When we read online, we do not read horizontally, line by line, but slide vertically along the text.

4. Identification. In communication on the Internet, there are two sorts of identifications: mental and physical connection with media that helps us get inside cyberspace or mental identification with content in cyberspace. The first type of identification constitutes mental or mentally physical cyborgism. It is currently possible to connect technology (artificial arm) to the nervous system and control it by thought. We can expect similar applications also in the field of information technology—for example, Google's Google Glass and later possibly electronic contact lenses. Some technologies may, in the future, be implemented also in the human body. This could bring its positives for some people who suffer from injuries after accidents and also provide immediate access to information. On the other hand, it could bring a fundamental dependence on technology and potential danger of abusing this technology to spy on people or control them.

At the present time, self-identification with content in cyberspace through social groups, or avatars, is still more and more common. The effort to find one's place among a social group and be able to share one's knowledge and experiences may be taken as desirable. One can sometimes feel the need to live a better life in cyberspace, for example, in a videogame called Second life. This can induce therapeutic, liberating effects. People can feel a need to become somebody else in life and demonstrate this also in social life, as we can see, for example, in a videogame called cosplay (a portmanteau of the words costume play). In Japan, but now also in other counties in the Western world, cartoons and cartoon characters are idols for teenagers. Young people identify themselves with these characters, which manifest the most in their costumes. Sometimes this new identity is so strong that young people will not want to abandon the idea [9]. Everything depends on the extent and manifestation of such identification. If it causes alienation or addiction, it becomes a negative situation.

Analysis of positive and negative changes in communication in cyberspace reveals that we need media education. D. Petranová [23] explains it is critical thinking that is the most important objective, and this can help us treat media with reserve, analyze information correctly, think independently, free ourselves from stereotypes, and so on. This all should improve our personal freedom.

5. Conclusion

We have known communication in cyberspace, especially in the cyberspace Internet, for slightly over a generation span, and we can already say that it has significantly influenced our cultural and social life; it even initiated a new existential dimension. The Internet cyberspace is a medium through which we create our ideas, communicate, and learn. Basing on analysis of older types of media, for example, written word and printed word, we know that these managed to restructure human thinking and acting completely. This leads us to believe that something similar is happening, and will be happening, also in connection with the Internet cyberspace. Media, including the Internet, influence us simply because we use them. The mere fact that we are connected to the Internet and use it in our communication in cyberspace is all what it takes; how we use it is not so important. We call the first type of influence, which is the result of being connected to technology, formal influence. The second kind of influence, triggered by communicating certain content, is defined as content influence. In this article we tried to point out that formal type changes are more crucial and paradigmatical and even constitute a new anthropogenesis. We specifically studied changes in our ideas of time and space, structure of thinking, and identity in cyberspace. These changes do not manifest merely in communication in cyberspace but affect also our everyday life. This is the reason why it is necessary to know their scope, positives, and negatives. New communication technologies influence our mentality but also our physical body. The question is how much is just enough to refine our personality, knowledge, and freedom and how much is simply too much, so they will start dictating and conducting us. We therefore need to learn to trust media with reserve, be critical, and spend at least part of the time we have without the influence of media, especially away from the Internet cyberspace.

Author details

Slavomír Gálik* and Sabína Gáliková Tolnaiová
Faculty of Mass Media Communication, University of Ss. Cyril and Methodius in Trnava, Slovakia

*Address all correspondence to: s_galik@yahoo.com

References

[1] Clark D. Characterizing Cyberspace: Past, Present and Future [Online]. 2010. Available from: http://web.mit.edu/ecir/pdf/clark-cyberspace.pdf [Accessed: 15 December 2013]

[2] Holmes D. Communication Theory. Media, Technology and Society. London: Sage; 2010. p. 243

[3] Cocking D. Plural selves and relational identity. Intimacy and privacy online. In: Van de Hoven J, Weckert J, editors. Information Technology and Moral Philosophy. Cambridge: University Press New York; 2009. pp. 123-141

[4] Sprondel J, Breyer T, Wehrle M. Cyber Anthropology Being Human on the Internet. [Online]. 2011. Available from: http://www.hiig.de/wp-content/uploads/2012/04/CyberAnthropology-Paper.pdf [Accessed: 16 November 2014]

[5] Ropolyi L. Philosophy of the Internet. A Discourse on the Nature of the Internet. [Online]. 2013. Available from: http://elte.prompt.hu/sites/default/files/tananyagok/philosophy_of_internet/book.pdf [Accessed: 06 January 2015]

[6] Růžička M. Informace a Dobro [Information and Good]. Praha: Ježek; 1993. p. 84

[7] Cejpek J. Informace, Komunikace a Myšlení. Úvod do Informační Vědy [Information, Communication and Thinking. Introduction to Information Science]. Praha: Karolinum; 2005. p. 234

[8] Lohisse J. Komunikační systémy [Communication Systems]. Praha: Karolinum; 2003. p. 200

[9] Gálik S. Filozofia a médiá [Philosophy and Media]. Bratislava: Iris; 2012. p. 104

[10] Rankov P. Informačná Spoločnosť—Perspektívy, Problémy, Paradoxy [Information Society—Perspectives, Problems, Paradoxes]. Levice: LCA Publisher Group; 2006. p. 175

[11] Postman N. Ubavit se k Smrti. Veřejná Komunikace Ve věku zábavy [Amusing Ourselves to Death: Public Discourse in the Age of Show Business]. Praha: Mladá fronta; 2010. p. 208

[12] Pravdová H. Determinanty Kreovania mediálnej kultúry [Determinants of Creating Media Culture]. Trnava: FMK UCM; 2009. p. 361

[13] Sartori G. Homo Videns: La Sociedad Teledirigada [Online]. 1997. Available from: http://ifdc6m.juj.infd.edu.ar/aula/archivos/repositorio/0/116/HOMO_VIDENS.pdf [Accessed: 15 December 2013]

[14] Cenká R, Lužák I. Kyberpriestor a fenomén mystickej smrti [the cyberspace and the phenomenon of "mystical death"]. In: Gálik S, editor. K Problému Univerzálnosti a Aktuálnosti Fenoménu Mystickej Smrti. Łódź: KSIĘŻY MŁYN Dom Wydawniczy Michał Koliński; 2013. pp. 227-262

[15] Gáliková Tolnaiová S. Anthropological risks and the form that evil takes in the electronic media era. In: Jozek M, editor. Contemporary Images of Evil. Krakow: Wydawnictwo Naukowe Uniwersytetu Pedagogicznego; 2013. pp. 33-56

[16] Bauerlein M. Najhlúpejšia Generácia. Ako Digitálna éra Ohlupuje Mladých Američanov a Ohrozuje Našu Budúcnosť [The Dumbest Generation: How the Digital Age Stupefies Young Americans and Jeopardizes our Future]. Bratislava: Vydavateľstvo Spolku slovenských spisovateľov; 2010. p. 208

[17] Volko L. Internetová komunikácia ako súčasť mediálnej kultúry [internet communication as part of media culture]. In: Magál S, Mikuš T, Petranová D, editors. Megatrendy a Médiá. Limity Mediálnej Internetovej komunikácie. Trnava: FMK UCM; 2011. pp. 87-96

[18] Bystřický J. Média, Komunikace a Kultura [Media, Communication and Culture]. Plzeň: Aleš Čeněk; 2008. p. 96

[19] Solík, M. Semiotic approach to analysis of advertising. In: European Journal of Science and Theology. 2014;**10**(1):207-217

[20] Eco U. Od Stromu k Labyrintu. Historické Studie o Znaku a Interpretaci [from Tree to Labyrinth. Historical Studies of Sign and Interpretation]. Praha: Argo; 2012. p. 653

[21] Spitzer M. Digitální Demence [Digital Dementia]. Brno: Host; 2014. p. 343

[22] Višňovský J. Aktuálne otázky teórie a Praxe žurnalistiky v ére Internetu [Current Issues of Journalism Theory and Practice in the Internet Era]. Trnava: FMK UCM; 2015. p. 350

[23] Petranová D. Mediálna výchova a kritické Myslenie [Media Education and Critical Thinking]. Trnava: FMK UCM; 2013. p. 108

Chapter 3

Research Design and Methodology

Kassu Jilcha Sileyew

Abstract

There are a number of approaches used in this research method design. The purpose of this chapter is to design the methodology of the research approach through mixed types of research techniques. The research approach also supports the researcher on how to come across the research result findings. In this chapter, the general design of the research and the methods used for data collection are explained in detail. It includes three main parts. The first part gives a highlight about the dissertation design. The second part discusses about qualitative and quantitative data collection methods. The last part illustrates the general research framework. The purpose of this section is to indicate how the research was conducted throughout the study periods.

Keywords: research design, methodology, sampling, data sources, population, workplace

1. Introduction

Research methodology is the path through which researchers need to conduct their research. It shows the path through which these researchers formulate their problem and objective and present their result from the data obtained during the study period. This research design and methodology chapter also shows how the research outcome at the end will be obtained in line with meeting the objective of the study. This chapter hence discusses the research methods that were used during the research process. It includes the research methodology of the study from the research strategy to the result dissemination. For emphasis, in this chapter, the author outlines the research strategy, research design, research methodology, the study area, data sources such as primary data sources and secondary data, population consideration and sample size determination such as questionnaires sample size determination and workplace site exposure measurement sample determination, data collection methods like primary data collection methods including workplace site observation data collection and data collection through desk review, data collection through questionnaires, data obtained from experts opinion, workplace site exposure measurement, data collection tools pretest, secondary data collection methods, methods of data analysis used such as quantitative data analysis and qualitative data analysis, data analysis software, the reliability and validity analysis of the quantitative data, reliability of data, reliability analysis, validity, data quality management, inclusion criteria, ethical consideration and dissemination of result and its utilization approaches. In order to satisfy the objectives of the study, a qualitative and quantitative research method is apprehended in general. The study used these mixed strategies because the data were obtained from all aspects of the data source during the study time. Therefore, the purpose of this methodology is to satisfy the research plan and target devised by the researcher.

2. Research design

The research design is intended to provide an appropriate framework for a study. A very significant decision in research design process is the choice to be made regarding research approach since it determines how relevant information for a study will be obtained; however, the research design process involves many interrelated decisions [1].

This study employed a mixed type of methods. The first part of the study consisted of a series of well-structured questionnaires (for management, employee's representatives, and technician of industries) and semi-structured interviews with key stakeholders (government bodies, ministries, and industries) in participating organizations. The other design used is an interview of employees to know how they feel about safety and health of their workplace, and field observation at the selected industrial sites was undertaken.

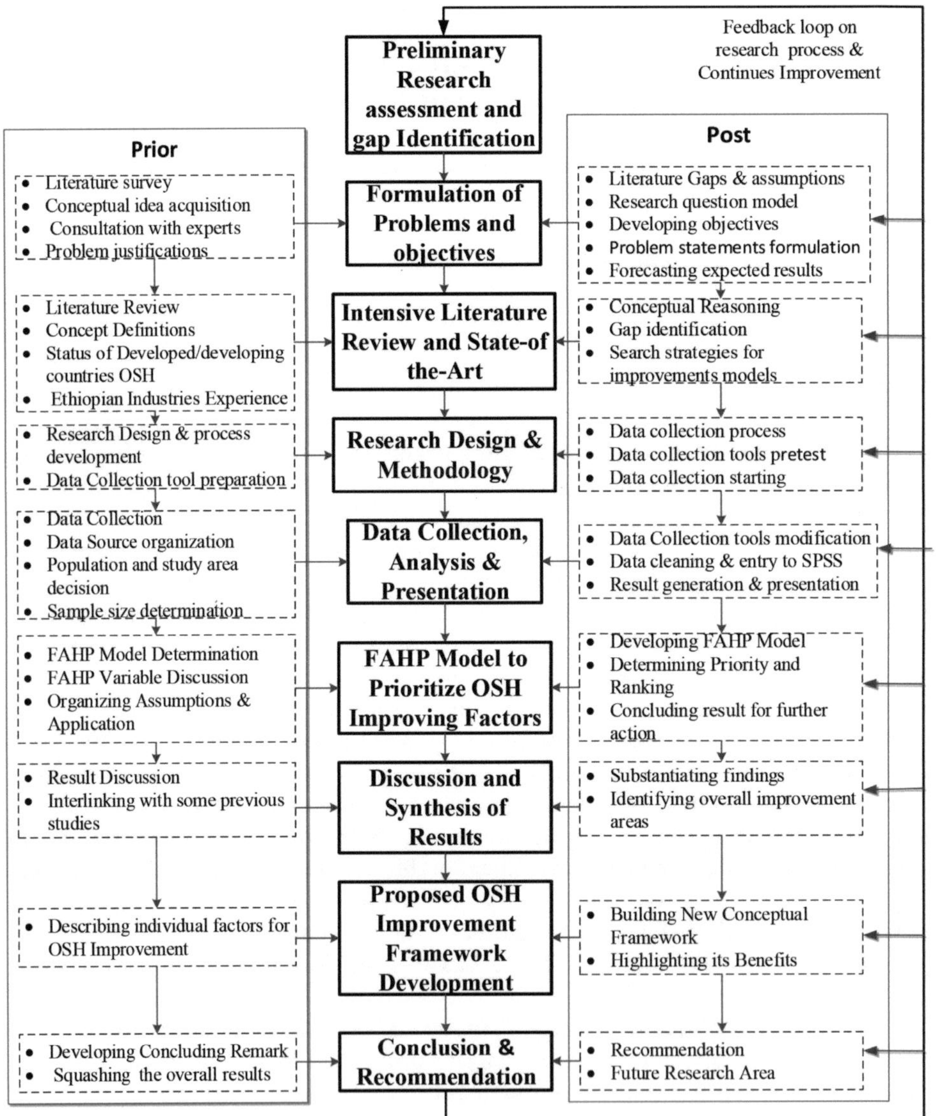

Figure 1.
Research methods and processes (author design).

Hence, this study employs a descriptive research design to agree on the effects of occupational safety and health management system on employee health, safety, and property damage for selected manufacturing industries. Saunders et al. [2] and Miller [3] say that descriptive research portrays an accurate profile of persons, events, or situations. This design offers to the researchers a profile of described relevant aspects of the phenomena of interest from an individual, organizational, and industry-oriented perspective. Therefore, this research design enabled the researchers to gather data from a wide range of respondents on the impact of safety and health on manufacturing industries in Ethiopia. And this helped in analyzing the response obtained on how it affects the manufacturing industries' workplace safety and health. The research overall design and flow process are depicted in **Figure 1**.

3. Research methodology

To address the key research objectives, this research used both qualitative and quantitative methods and combination of primary and secondary sources. The qualitative data supports the quantitative data analysis and results. The result obtained is triangulated since the researcher utilized the qualitative and quantitative data types in the data analysis. The study area, data sources, and sampling techniques were discussed under this section.

3.1 The study area

According to Fraenkel and Warren [4] studies, population refers to the complete set of individuals (subjects or events) having common characteristics in which the researcher is interested. The population of the study was determined based on random sampling system. This data collection was conducted from March 07, 2015 to December 10, 2016, from selected manufacturing industries found in Addis Ababa city and around. The manufacturing companies were selected based on their employee number, established year, and the potential accidents prevailing and the manufacturing industry type even though all criterions were difficult to satisfy.

3.2 Data sources

3.2.1 Primary data sources

It was obtained from the original source of information. The primary data were more reliable and have more confidence level of decision-making with the trusted analysis having direct intact with occurrence of the events. The primary data sources are industries' working environment (through observation, pictures, and photograph) and industry employees (management and bottom workers) (interview, questionnaires and discussions).

3.2.2 Secondary data

Desk review has been conducted to collect data from various secondary sources. This includes reports and project documents at each manufacturing sectors (more on medium and large level). Secondary data sources have been obtained from literatures regarding OSH, and the remaining data were from the companies' manuals, reports, and some management documents which were included under the desk review. Reputable journals, books, different articles, periodicals, proceedings, magazines, newsletters, newspapers, websites, and other sources were considered on the manufacturing industrial sectors. The data also obtained from the existing working documents, manuals, procedures, reports, statistical data, policies, regulations, and standards were taken into account for the review.

In general, for this research study, the desk review has been completed to this end, and it had been polished and modified upon manuals and documents obtained from the selected companies.

4. Population and sample size

4.1 Population

The study population consisted of manufacturing industries' employees in Addis Ababa city and around as there are more representative manufacturing industrial clusters found. To select representative manufacturing industrial sector population, the types of the industries expected were more potential to accidents based on random and purposive sampling considered. The population of data was from textile, leather, metal, chemicals, and food manufacturing industries. A total of 189 sample sizes of industries responded to the questionnaire survey from the priority areas of the government. Random sample sizes and disproportionate methods were used, and 80 from wood, metal, and iron works; 30 from food, beverage, and tobacco products; 50 from leather, textile, and garments; 20 from chemical and chemical products; and 9 from other remaining 9 clusters of manufacturing industries responded.

4.2 Questionnaire sample size determination

A simple random sampling and purposive sampling methods were used to select the representative manufacturing industries and respondents for the study. The simple random sampling ensures that each member of the population has an equal chance for the selection or the chance of getting a response which can be more than equal to the chance depending on the data analysis justification. Sample size determination procedure was used to get optimum and reasonable information. In this study, both probability (simple random sampling) and nonprobability (convenience, quota, purposive, and judgmental) sampling methods were used as the nature of the industries are varied. This is because of the characteristics of data sources which permitted the researchers to follow the multi-methods. This helps the analysis to triangulate the data obtained and increase the reliability of the research outcome and its decision. The companies' establishment time and its engagement in operation, the number of employees and the proportion it has, the owner types (government and private), type of manufacturing industry/production, types of resource used at work, and the location it is found in the city and around were some of the criteria for the selections.

The determination of the sample size was adopted from Daniel [5] and Cochran [6] formula. The formula used was for unknown population size Eq. (1) and is given as

$$n = \frac{Z^2 P(1-P)}{d^2} \tag{1}$$

where n = sample size, Z = statistic for a level of confidence, P = expected prevalence or proportion (in proportion of one; if 50%, P = 0.5), and d = precision (in proportion of one; if 6%, d = 0.06). Z statistic (Z): for the level of confidence of 95%, which is conventional, Z value is 1.96. In this study, investigators present their results with 95% confidence intervals (CI).

The expected sample number was 267 at the marginal error of 6% for 95% confidence interval of manufacturing industries. However, the collected data indicated that only 189 populations were used for the analysis after rejecting some data having more missing values in the responses from the industries. Hence, the actual data collection resulted in 71% response rate. The 267 population were assumed to be satisfactory and representative for the data analysis.

4.3 Workplace site exposure measurement sample determination

The sample size for the experimental exposure measurements of physical work environment has been considered based on the physical data prepared for questionnaires and respondents. The response of positive were considered for exposure measurement factors to be considered for the physical environment health and disease causing such as noise intensity, light intensity, pressure/stress, vibration, temperature/coldness, or hotness and dust particles on 20 workplace sites. The selection method was using random sampling in line with purposive method. The measurement of the exposure factors was done in collaboration with Addis Ababa city Administration and Oromia Bureau of Labour and Social Affair (AACBOLSA). Some measuring instruments were obtained from the Addis Ababa city and Oromia Bureau of Labour and Social Affair.

5. Data collection methods

Data collection methods were focused on the followings basic techniques. These included secondary and primary data collections focusing on both qualitative and quantitative data as defined in the previous section. The data collection mechanisms are devised and prepared with their proper procedures.

5.1 Primary data collection methods

Primary data sources are qualitative and quantitative. The qualitative sources are field observation, interview, and informal discussions, while that of quantitative data sources are survey questionnaires and interview questions. The next sections elaborate how the data were obtained from the primary sources.

5.1.1 Workplace site observation data collection

Observation is an important aspect of science. Observation is tightly connected to data collection, and there are different sources for this: documentation, archival records, interviews, direct observations, and participant observations. Observational research findings are considered strong in validity because the researcher is able to collect a depth of information about a particular behavior. In this dissertation, the researchers used observation method as one tool for collecting information and data before questionnaire design and after the start of research too. The researcher made more than 20 specific observations of manufacturing industries in the study areas. During the observations, it found a deeper understanding of the working environment and the different sections in the production system and OSH practices.

5.1.2 Data collection through interview

Interview is a loosely structured qualitative in-depth interview with people who are considered to be particularly knowledgeable about the topic of interest.

The semi-structured interview is usually conducted in a face-to-face setting which permits the researcher to seek new insights, ask questions, and assess phenomena in different perspectives. It let the researcher to know the in-depth of the present working environment influential factors and consequences. It has provided opportunities for refining data collection efforts and examining specialized systems or processes. It was used when the researcher faces written records or published document limitation or wanted to triangulate the data obtained from other primary and secondary data sources.

This dissertation is also conducted with a qualitative approach and conducting interviews. The advantage of using interviews as a method is that it allows respondents to raise issues that the interviewer may not have expected. All interviews with employees, management, and technicians were conducted by the corresponding researcher, on a face-to-face basis at workplace. All interviews were recorded and transcribed.

5.1.3 Data collection through questionnaires

The main tool for gaining primary information in practical research is questionnaires, due to the fact that the researcher can decide on the sample and the types of questions to be asked [2].

In this dissertation, each respondent is requested to reply to an identical list of questions mixed so that biasness was prevented. Initially the questionnaire design was coded and mixed up from specific topic based on uniform structures. Consequently, the questionnaire produced valuable data which was required to achieve the dissertation objectives.

The questionnaires developed were based on a five-item Likert scale. Responses were given to each statement using a five-point Likert-type scale, for which 1 = "strongly disagree" to 5 = "strongly agree." The responses were summed up to produce a score for the measures.

5.1.4 Data obtained from experts' opinion

The data was also obtained from the expert's opinion related to the comparison of the knowledge, management, collaboration, and technology utilization including their sub-factors. The data obtained in this way was used for prioritization and decision-making of OSH, improving factor priority. The prioritization of the factors was using Saaty scales (1–9) and then converting to Fuzzy set values obtained from previous researches using triangular fuzzy set [7].

5.1.5 Workplace site exposure measurement

The researcher has measured the workplace environment for dust, vibration, heat, pressure, light, and noise to know how much is the level of each variable. The primary data sources planned and an actual coverage has been compared as shown in **Table 1**.

The response rate for the proposed data source was good, and the pilot test also proved the reliability of questionnaires. Interview/discussion resulted in 87% of responses among the respondents; the survey questionnaire response rate obtained was 71%, and the field observation response rate was 90% for the whole data analysis process. Hence, the data organization quality level has not been compromised.

This response rate is considered to be representative of studies of organizations. As the study agrees on the response rate to be 30%, it is considered acceptable [8]. Saunders et al. [2] argued that the questionnaire with a scale response of 20%

Instrument	Planned	Actual coverage	Success level
Interview/discussion	15	13	87%
Survey questionnaires	267	189	71%
Observation	20	18	90%
Workplace site exposure measurement	20	20	100%

Table 1.
Planned versus actual coverage of the survey.

response rate is acceptable. Low response rate should not discourage the researchers, because a great deal of published research work also achieves low response rate. Hence, the response rate of this study is acceptable and very good for the purpose of meeting the study objectives.

5.1.6 Data collection tool pretest

The pretest for questionnaires, interviews, and tools were conducted to validate that the tool content is valid or not in the sense of the respondents' understanding. Hence, content validity (in which the questions are answered to the target without excluding important points), internal validity (in which the questions raised answer the outcomes of researchers' target), and external validity (in which the result can generalize to all the population from the survey sample population) were reflected. It has been proved with this pilot test prior to the start of the basic data collections. Following feedback process, a few minor changes were made to the originally designed data collect tools. The pilot test made for the questionnaire test was on 10 sample sizes selected randomly from the target sectors and experts.

5.2 Secondary data collection methods

The secondary data refers to data that was collected by someone other than the user. This data source gives insights of the research area of the current state-of-the-art method. It also makes some sort of research gap that needs to be filled by the researcher. This secondary data sources could be internal and external data sources of information that may cover a wide range of areas.

Literature/desk review and industry documents and reports: To achieve the dissertation's objectives, the researcher has conducted excessive document review and reports of the companies in both online and offline modes. From a methodological point of view, literature reviews can be comprehended as content analysis, where quantitative and qualitative aspects are mixed to assess structural (descriptive) as well as content criteria.

A literature search was conducted using the database sources like MEDLINE; Emerald; Taylor and Francis publications; EMBASE (medical literature); PsycINFO (psychological literature); Sociological Abstracts (sociological literature); accident prevention journals; US Statistics of Labor, European Safety and Health database; ABI Inform; Business Source Premier (business/management literature); EconLit (economic literature); Social Service Abstracts (social work and social service literature); and other related materials. The search strategy was focused on articles or reports that measure one or more of the dimensions within the research OSH model framework. This search strategy was based on a framework and measurement filter strategy developed by the Consensus-Based Standards for the Selection of Health Measurement

Instruments (COSMIN) group. Based on screening, unrelated articles to the research model and objectives were excluded. Prior to screening, researcher (principal investigator) reviewed a sample of more than 2000 articles, websites, reports, and guidelines to determine whether they should be included for further review or reject. Discrepancies were thoroughly identified and resolved before the review of the main group of more than 300 articles commenced. After excluding the articles based on the title, keywords, and abstract, the remaining articles were reviewed in detail, and the information was extracted on the instrument that was used to assess the dimension of research interest. A complete list of items was then collated within each research targets or objectives and reviewed to identify any missing elements.

6. Methods of data analysis

Data analysis method follows the procedures listed under the following sections. The data analysis part answered the basic questions raised in the problem statement. The detailed analysis of the developed and developing countries' experiences on OSH regarding manufacturing industries was analyzed, discussed, compared and contrasted, and synthesized.

6.1 Quantitative data analysis

Quantitative data were obtained from primary and secondary data discussed above in this chapter. This data analysis was based on their data type using Excel, SPSS 20.0, Office Word format, and other tools. This data analysis focuses on numerical/quantitative data analysis.

Before analysis, data coding of responses and analysis were made. In order to analyze the data obtained easily, the data were coded to SPSS 20.0 software as the data obtained from questionnaires. This task involved identifying, classifying, and assigning a numeric or character symbol to data, which was done in only one way pre-coded [9, 10]. In this study, all of the responses were pre-coded. They were taken from the list of responses, a number of corresponding to a particular selection was given. This process was applied to every earlier question that needed this treatment. Upon completion, the data were then entered to a statistical analysis software package, SPSS version 20.0 on Windows 10 for the next steps.

Under the data analysis, exploration of data has been made with descriptive statistics and graphical analysis. The analysis included exploring the relationship between variables and comparing groups how they affect each other. This has been done using cross tabulation/chi square, correlation, and factor analysis and using nonparametric statistic.

6.2 Qualitative data analysis

Qualitative data analysis used for triangulation of the quantitative data analysis. The interview, observation, and report records were used to support the findings. The analysis has been incorporated with the quantitative discussion results in the data analysis parts.

6.3 Data analysis software

The data were entered using SPSS 20.0 on Windows 10 and analyzed. The analysis supported with SPSS software much contributed to the finding. It had contributed to the data validation and correctness of the SPSS results. The software analyzed and compared the results of different variables used in the research questionnaires. Excel is also used to draw the pictures and calculate some analytical solutions.

7. The reliability and validity analysis of the quantitative data

7.1 Reliability of data

The reliability of measurements specifies the amount to which it is without bias (error free) and hence ensures consistent measurement across time and across the various items in the instrument [8]. In reliability analysis, it has been checked for the stability and consistency of the data. In the case of reliability analysis, the researcher checked the accuracy and precision of the procedure of measurement. Reliability has numerous definitions and approaches, but in several environments, the concept comes to be consistent [8]. The measurement fulfills the requirements of reliability when it produces consistent results during data analysis procedure. The reliability is determined through Cranach's alpha as shown in **Table 2**.

7.2 Reliability analysis

Cronbach's alpha is a measure of internal consistency, i.e., how closely related a set of items are as a group [11]. It is considered to be a measure of scale reliability. The reliability of internal consistency most of the time is measured based on the Cronbach's alpha value. Reliability coefficient of 0.70 and above is considered "acceptable" in most research situations [12]. In this study, reliability analysis for internal consistency of Likert-scale measurement after deleting 13 items was found similar; the reliability coefficients were found for 76 items were 0.964 and for the individual groupings made shown in **Table 2**. It was also found internally consistent using the Cronbach's alpha test. **Table 2** shows the internal consistency of the seven major instruments in which their reliability falls in the acceptable range for this research.

s/n	Qualitative data major groups	Items number	Alpha standardized)
1	Knowledge related factors	K01 to K08	0.864
2	Management related factors	M01 to M17	0.877
3	Technology and suppliers related factors	T01 to T10	0.792
4	Collaboration and support related factors	C01 to C07	0.781
5	Policy , standards and guidelines related factors	P01 to P08	0.888
6	Hazards and accidents related factors	H01 to H10	0.720
7	Personal Protective Equipment related factors	PPE01 to PPE10	0.931
	Total	70	0.966

K stands for knowledge; M, management; T, technology; C, collaboration; P, policy, standards, and regulation; H, hazards and accident conditions; PPE, personal protective equipment.

Table 2.
Internal consistency and reliability test of questionnaires items.

7.3 Validity

Face validity used as defined by Babbie [13] is an indicator that makes it seem a reasonable measure of some variables, and it is the subjective judgment that the instrument measures what it intends to measure in terms of relevance [14]. Thus, the researcher ensured, in this study, when developing the instruments that uncertainties were eliminated by using appropriate words and concepts in order to enhance clarity and general suitability [14]. Furthermore, the researcher submitted the instruments to the research supervisor and the joint supervisor who are both occupational health experts, to ensure validity of the measuring instruments and determine whether the instruments could be considered valid on face value.

In this study, the researcher was guided by reviewed literature related to compliance with the occupational health and safety conditions and data collection methods before he could develop the measuring instruments. In addition, the pretest study that was conducted prior to the main study assisted the researcher to avoid uncertainties of the contents in the data collection measuring instruments. A thorough inspection of the measuring instruments by the statistician and the researcher's supervisor and joint experts, to ensure that all concepts pertaining to the study were included, ensured that the instruments were enriched.

8. Data quality management

Insight has been given to the data collectors on how to approach companies, and many of the questionnaires were distributed through MSc students at Addis Ababa Institute of Technology (AAiT) and manufacturing industries' experience experts. This made the data quality reliable as it has been continually discussed with them. Pretesting for questionnaire was done on 10 workers to assure the quality of the data and for improvement of data collection tools. Supervision during data collection was done to understand how the data collectors are handling the questionnaire, and each filled questionnaires was checked for its completeness, accuracy, clarity, and consistency on a daily basis either face-to-face or by phone/email. The data expected in poor quality were rejected out of the acting during the screening time. Among planned 267 questionnaires, 189 were responded back. Finally, it was analyzed by the principal investigator.

9. Inclusion criteria

The data were collected from the company representative with the knowledge of OSH. Articles written in English and Amharic were included in this study. Database information obtained in relation to articles and those who have OSH area such as interventions method, method of accident identification, impact of occupational accidents, types of occupational injuries/disease, and impact of occupational accidents, and disease on productivity and costs of company and have used at least one form of feedback mechanism. No specific time period was chosen in order to access all available published papers. The questionnaire statements which are similar in the questionnaire have been rejected from the data analysis.

10. Ethical consideration

Ethical clearance was obtained from the School of Mechanical and Industrial Engineering, Institute of Technology, Addis Ababa University. Official letters were written from the School of Mechanical and Industrial Engineering to the respective manufacturing industries. The purpose of the study was explained to the study subjects. The study subjects were told that the information they provided was kept confidential and that their identities would not be revealed in association with the information they provided. Informed consent was secured from each participant. For bad working environment assessment findings, feedback will be given to all manufacturing industries involved in the study. There is a plan to give a copy of the result to the respective study manufacturing industries' and ministries' offices. The respondents' privacy and their responses were not individually analyzed and included in the report.

11. Dissemination and utilization of the result

The result of this study will be presented to the Addis Ababa University, AAiT, School of Mechanical and Industrial Engineering. It will also be communicated to the Ethiopian manufacturing industries, Ministry of Labor and Social Affair, Ministry of Industry, and Ministry of Health from where the data was collected. The result will also be availed by publication and online presentation in Google Scholars. To this end, about five articles were published and disseminated to the whole world.

12. Conclusion

The research methodology and design indicated overall process of the flow of the research for the given study. The data sources and data collection methods were used. The overall research strategies and framework are indicated in this research process from problem formulation to problem validation including all the parameters. It has laid some foundation and how research methodology is devised and framed for researchers. This means, it helps researchers to consider it as one of the samples and models for the research data collection and process from the beginning of the problem statement to the research finding. Especially, this research flow helps new researchers to the research environment and methodology in particular.

Conflict of interest

There is no "conflict of interest."

Author details

Kassu Jilcha Sileyew
School of Mechanical and Industrial Engineering, Addis Ababa Institute of Technology, Addis Ababa University, Addis Ababa, Ethiopia

*Address all correspondence to: jkassu@gmail.com

References

[1] Aaker A, Kumar VD, George S. Marketing Research. New York: John Wiley & Sons Inc; 2000

[2] Saunders M, Lewis P, Thornhill A. Research Methods for Business Student. 5th ed. Edinburgh Gate: Pearson Education Limited; 2009

[3] Miller P. Motivation in the Workplace. Work and Organizational Psychology. Oxford: Blackwell Publishers; 1991

[4] Fraenkel FJ, Warren NE. How to Design and Evaluate Research in Education. 4th ed. New York: McGraw-Hill; 2002

[5] Danniel WW. Biostatist: A Foundation for Analysis in the Health Science. 7th ed. New York: John Wiley & Sons; 1999

[6] Cochran WG. Sampling Techniques. 3rd ed. New York: John Wiley & Sons; 1977

[7] Saaty TL. The Analytical Hierarchy Process. Pittsburg: PWS Publications; 1990

[8] Sekaran U, Bougie R. Research Methods for Business: A Skill Building Approach. 5th ed. New Delhi: John Wiley & Sons, Ltd; 2010. pp. 1-468

[9] Luck DJ, Rubin RS. Marketing Research. 7th ed. New Jersey: Prentice-Hall International; 1987

[10] Wong TC. Marketing Research. Oxford, UK: Butterworth-Heinemann; 1999

[11] Cronbach LJ. Coefficient alpha and the internal structure of tests. Psychometrika. 1951;**16**:297-334

[12] Tavakol M, Dennick R. Making sense of Cronbach's alpha. International Journal of Medical Education. 2011;**2**: 53-55. DOI: 10.5116/ijme.4dfb.8dfd

[13] Babbie E. The Practice of Social Research. 12th ed. Belmont, CA: Wadsworth; 2010

[14] Polit DF, Beck CT. Generating and Assessing Evidence for Nursing Practice. 8th ed. Williams and Wilkins: Lippincott; 2008

Chapter 4

Cyberspace as a New Living World and Its Axiological Contexts

Sabína Gáliková Tolnaiová and Slavomír Gálik

Abstract

The subject of the chapter is cyberspace in an axiological perspective, which is our new lifeworld. The focus is particularly on the problem of the quality of our life in its specific circumstances. The aim is (on the background of the characteristics of cyberspace as a lifeworld) to solve the problem of values and significance, but also the risks of our so-called cyber experience. In this context, the aim is also to identify various conditions, axiological indicators and the relevant elements of the quality of our life in cyberspace. The authors pursue their goal using the phenomenological-hermeneutic method within the four parts of the chapter. In part 1, cyberspace is interpreted as a life world that is co-constructed in our acts of communication. In part 2, the problem of values, significance and risks of our cyber experience is discussed. The key variable is digital "well-being." As they point out in part 3, it should be our morally based value "good life," which is expressed as "ethos" in our life. In part 4, in this perspective, we are faced with the relevant task of the art of living 'ars vivendi' with the necessary coherent self-understanding and value-moral claims and the education should also have a "psychological" dimension.

Keywords: internet, cyberspace, living world, values, well-being, good life, ars vivendi, education, psychagogy

1. Introduction

New digital technologies, or media, have brought us a new phenomenon-the so-called cyberspace. Once we enter cyberspace and communicate in it, it is not just our communication space that expands, the same happens with our living space.

Though it is obvious that the internet brings also another, alternative form of cyberspace[1], it is now most frequently associated specifically with the internet

[1] We should keep in mind that the internet is not the only network, as even before the internet there was a lengthy development of networks that actually qualified as domains of cyberspace [1]. On the other hand, similar to the past, also now different forms of cyberspace develop parallelly in different ways (e.g., Clark mentions various connections between the four levels of cyberspace). However, we can say that their structures and structural implications are not very different [2].

[2, 3]. It is understood to be the most dominant place for (social) communication. As Cappuro points out, individual people are connected with each other through global communication. Cyberspace as such allows various synergies inside and outside our political, ethnical, economical and cultural boundaries or differences [4].

The aim of the chapter is (on the background of the characteristics of cyberspace as a life world) to solve the problem of values and significance, but also the risks of our so-called cyber experience, in this context, to identify various conditions, axiological indicators and the relevant elements of the quality of our life in cyberspace itself.

In the following chapters by the phenomenological-hermeneutic method, we will be talking about the internet cyberspace as of a new, specific living world, in which we spend a great deal of time to create our identity, and where we live our specific cyber experience. In this context, we will focus especially on its axiological dimension. We will also look at some values and risks that come along. Further, we will examine digital welfare (well-being) and "good life" in the cyberspace of internet in the axiological and ethical aspect. We will also speak of the virtue to live "ars vivendi" and, finally, about education that can help us achieve this goal.

2. Cyberspace of the internet as our living world

The term cyberspace generally describes an interface between computers and people, or a meeting point for digital information and human perception. However, it is also often used to refer to interaction between people using computers, especially through the internet [3]. We define the internet and other computer networks as collective usage of virtual entities [5]. Thanks to its new ways, the internet and other computer networks have introduced a change in the nature of social interaction, or communication, new possibilities and routines. Its net-like nature and structure have contributed to expansion of mutual space and, as Nanni points out, this form of media is able to start interaction between masses and wipe out territories [6].[2]

We can state here that the internet cyberspace is for people who are not solely its passive users simply because these people actually co-create it. They define and shape its character and actively create its content through the way they use it. This is the reason why people are the most important component, the highest level [2]. Cyberspace thus represents a kind of socially constructed world or dimension, an electronic Agora-a central public space. It is a "cyber-café" [1]. Also Hakken, as Macek points out, understands cyberspace as a social arena, place for social interaction between those who use advanced communication and information technology. This definition covers any and every possible lifestyle that is bound up with cultural existence mediated by this advanced technology [7].

It is possible to state that the internet cyberspace as such has become our new living world in which we communicate, learn, do business or get entertained. It

[2] As a form of media, the internet can enable access, contact, exchange and discussion in an enhanced connection with every corner of the world, where a terminal is connected to the network. It seems it can support pluralism as well as unification; digital culture is destructuralised and decentralised. The internet world breaks institutional forms and disregards race hierarchy, gender and ethnicity. It subverts rational and logocentric forms of political authority [1].

is a specific place that reflects a vast part of our personal life, so in this context, we can perhaps mention that it becomes a place in which we like to spend some time-and feel almost at home. Cocking explains that computer technology offers a range of dimensions that we can use to express and develop our personal identity and various kinds of relationships. We can for example use the text-based email and chat-room, forums or web-site and web-cam technology to present ourselves, start professional, but also personal relationships and participate in any possible communities (based on hobby, interests and so on) [8].

Reflecting the fact that the internet cyberspace represents a great means for self-expression and communication in the present society [9] and, as Cocking points out, that we today present ourselves extensively when we conduct a great deal of activities and relationships with the help of computer technology, together with the author of this paper, we can ask what kind of online identity and relationship it is actually possible to create [8].

It reveals that cyberspace can be understood as a dimension that constitutes almost limitless possibilities for new forms of identity and behaviour [1]. Here we can develop, change or multiply our identity. Also Deuze mentions that in the world of media, we have an opportunity to create various versions of not only ourselves but also other people, and we are free to form and shape these versions at will. We can project, co-construct and bring to life one or various versions of ourselves in media. We can cooperate with other people to construct self-presentations and share them [10]. In this context, cyberspace is a sort of "screen" to show our dreams, desires and ideas. It is a form of extension for our creativity that helps us present ourselves. Similarly, Turkle [11] points out that computer acts as our new mirror that brings some influence, in it we consequently turn ourselves into objects and thus create our second nature.[3]

Discussions about our social interactions often emphasise difference between online and offline interactions, with cyberspace understood to be the distinctive place for such interactions [12]. However, shaping our identity here is not very different to shaping it in the "real" world. In fact, there are two mutual problems [9]. In cyberspace too, we deal with relations "world-person". It is interactions with other people that define what persons we become here; our identity is partially defined by what physical relations we have with other people. Relevant for cyberspace are consequences of our online interactions that, however, are influenced by our physical world. The physical nature is distinguishable in our online interactions and acts as a distinguishing factor for who we are, or what we mean for other people [13]. We can also notice that in cyberspace, our new living world, we can indeed have a multitude of partial identities, even simultaneously, but these are not independent from our subjective situation or social and cultural environment in "real" living world [9].

When a person is submerged in cyberspace, his or her experience is mediated; this person thus becomes a part of specific experience-the so called cyber experience. It appears that modern people like to spend time in this new

[3] As Cocking states also in this perspective, many theoretical approaches reveal cyberspace identity as much more fluid and variable. Cyberspace thus approaches this unstable, fractured and inconsistent "ME" and sees it as multiple identity. This "ME" is freed from physical world [8]. People can construct and shape this "ME" as something that is subject of numerous online versions of "ME", where for example gender does not play a role. We need to state here that these approaches mean, in fact, a denaturalised process of shaping the subject [12].

living world and that cyber experience as such is very attractive. As a next step, we will explore potential of values and risks that come with internet communication.

3. Cyber experience in the internet dimension, values and risks that come along

In the cyberspace living world, we cooperate with others, communicate and create (virtual) societies. Here, we are, as Deuze notices, more that ever interconnected medially with other people and we cannot overlook and disregard other people's lives [10]. We become part of a specific social cyber experience as intersubjectively communicating individuals. This cyber experience that becomes both individual and collective at the same time seems to be rather attractive for us. However, in what way does it become really valuable?

General good, a value hidden in humane communication, can generally be found when communication meets its mission, in other words, where it contributes to closeness, reciprocity and understanding between people, where it helps individuals manifest their feelings, intentions, where people look for mutual understanding, which is, as E. Višňovský points out, the nature of humane communication [14]. We can say that communication that favours intersubjectivity based on our own understanding of what is valuable and meaningful means general good for us. Such communication constitutes society that is based on mutual closeness and shared experience. As Rankov points out, communication on the internet brings satisfaction, positive feelings triggered by content that we communicate, but also by sharing other people's experience and communication as such [15]. Joy, happiness, pleasure and meaningful communication that the cyber experience of communication offers are a few of the qualities that we associate with "good life". However, is there a place for real mutual understanding, or its importance?

Višňovský notices that starting an interaction through the internet and finding mutual understanding and fellowship are two different things [14].[4] Also Bauman and Lyon warn in this context, when they say that rather than fellowship, we often find just the net itself. However, this net does not interpret true human community, because it does not look after us and is unreliable [16].[5] In this perspective, communicating community cannot win over offline non-network communication, and it seems that its cyber experience cannot compare to real human understanding that is relevant for healthy social bounds or society.

However, we can use also another approach to virtual communities, network communication and values that come with it. For example, Deuze explains that here social bounds are not based on mutual experiences or history, but chiefly on information exchange and talking about life. Here, sociability means lively and fleeting interactive social relationships, ephemeral but intense communication sessions. It is a specific kind of network sociability that seeks for contact and interaction, but also sensitive passionate and emotional communication and conversation. Offering emotional intimacy and credit, it is not meaningless. Deuze believes that societies

[4] According to Višňovský, one often looks for community, or communication, because it is needed but not found. In day-to-day life, we only find interaction, contact. The possibilities of virtual communication with the whole world are, Višňovský believes, utterly inadequate for us [14].

[5] Bauman and Lyon believe that the real, trusty community in sociological sense is reliable, our position here is more stable as we are confronted with duties and restrictions; we are watched and punished when community thinks it is necessary. Network, on the other side, lets us disconnect any time we want, so we are much freer to do what we like. Plus, and this is very important, network offers entertainment [16].

that use virtual dimension to communicate may constitute more fragile communities, but these are still meaningful and offer rather coherent understanding of the individual ME that should not be underestimated [10]. It is obvious that collectives of people who communicate in the internet cyberspace really constitute a totally new kind of society [15].

We should not overlook that it is the already mentioned sociability that makes new forms of media, especially the internet, attractive [10] and that it is also virtual communities that serve now as engine for flourishing and surprising life in this universal dimension that was born of contact [17]. However, this does not detract from the fact that the very cellular detachedness of individuals presents a risk for cyber experience within the network-based internet communication. Kováč and Gyén state that the internet, as a social technology used primarily for communication, can ignite social isolation in individual people, bring feelings of loneliness and depression and destroy their well-being [18]. Brožík notices that we witness people who become loners, people sitting in front of computer screens, losing grasp of their own life because their virtual partner actually drags them further and further from the real world [19]. Furthermore, let us notice for example anonymity, a specific determinant of communication in the internet cyberspace. As Rankov notices, along with neutralisation of social status, anonymity also triggers disinhibition, or unrestrained behaviour. We can on the one hand see that there is probably a connection with growing bravery to express ideas. We are more open, more sociable and capable of expressing what we think. We are less stiff and more inclined to joke and dare to be unique. On the other hand, however, feeling protected by anonymity, we are prone to breaking the norms, telling lies and being aggressive and vulgar when we deal with others [15]. Disinhibition in the context of relative anonymity and physical safety can hurt our self-confidence and favour intimacy that may open the gate for anger and hatred and thus make us aggressive and violent [18]. We can therefore state that anonymity and disinhibition pose individual and social risk in our cyber experience on the internet, which influences the whole value that it brings.

We should also notice the cyber experience of construction of ME, or our identity, online. As we have already mentioned, together with Deuze, we routinely create a vast variety of versions of ME on the internet. We project and develop one or a number of versions of ME, but to be more precise, we do this in cooperation with others when we constantly share these versions and self-presentations of ME [10]. We thus offer our self-image for others, even though not completely. We all, yet each of us individually, show what it means to be a human, what values are important to us, how we distinguish what is good and what is evil and what it actually means for us to be alive. We share our ideas and visions when we co-create and present our identity. We can say that whether intentionally or non-intentionally, this way we declare our meaning of life.

Obviously, our image on the internet depends on how we (intersubjectively, or publicly) self-project our position [13]. In the light of this, the internet, or communication using the internet, can surely also deconstruct our subjective identity, as, for example, Kuzior [20] notes. Bystřický warns that media reality as relation reality on the one hand brings a new paradigm into development stages of the extension of ME, but on the other hand, it also brings a risk of copying someone else's attitudes and ways of constructing a social, mental and aesthetic pattern. Relation-based character of media-presented reality may imply states of multiphrenia, individuals splitting to non-homogeneous segments, or multiply our own and private investments into empty and useless forms of self-presentations, false expression of hypothetical possibilities of one's own development [21]. In the context of cyber experience, we then face a risk of losing identity and depersonalisation, and this introduces a relevant question of interiority of the subject in cyberspace, or his or her coherence in time, which is necessary.

Who we became or are becoming in the internet cyberspace does make a difference. The way we deal with "life in media", using Deuze's words [10]-what we can and what we actually do invest in our relation to the others, is important. By our self-projection, self-construction and self-presentation in cyberspace, we become a part of collective process of "learning to live", in which intersubjectivity is bound to our understanding of what is relevant and valuable. In this perspective, our attitude is similar to that of Deuze [10]-we understand that it is reasonable to see self-expression of individuals today as more and more important in the cyberspace of internet.

According to Baeva, analysis of the nature of change of values in modern man reveals that rise of e-culture has led to construction of new values (electronic communication, e-spare time, e-creativity and so on) [22]. However, it is necessary to point out that cyber experience also brings certain risks presented by influence of the very technology, yet these risks cannot be specified in the reflection of cyber experience because they influence its value. We should not forget that also M. McLuhan speaks of self-amputation in connection with technological extension of man [23]. Similarly, also Bystřický points out that it seems we will pay for technological development by reducing one of the dimensions of our living world. Each new discovery in technology influences our personal living world and social system and imprints its own perspective onto the map of our individual and collective perception [21].

Therefore, to get the maximum advantage of cyber experience, it is important to be able to cope with various effects. Here, authors such as Gui et al. [24] define the so-called digital well-being.

4. Digital well-being and "good" life in the cyberspace of internet

The concept of digital well-being seen in Gui et al. is emerging right now, with communication stimuli overflow becoming hard to deal with [24]. Seeing how the internet is used now and how important it is for our communication and living world in the axiological context, we take this concept as undoubtedly relevant. These authors define and understand digital well-being as a state in which our subjective digital comfort is maintained by surplus of digital communication. In this state of well-being, individuals are able to use digital media to ensure their subjective comfort, safety, happiness and satisfaction. Such digital well-being secures general well-being of the subject in both hedonic and eudemonic perspective. It does not concentrate just on satisfaction and minimization of side effects of using digital media (hedonic dimension) but also on the ability to use this technology to present a meaningful help to one's own potential in life (eudemonic perspective) [24].

The way we handle digital media is, we believe, a key element for quality life. Theoretical approaches and empiric findings clearly identify a number of ways how media contents and media usage influence our everyday happiness, satisfaction with life, our effort to develop our personality and understand meaning of life, as Reinecke and Oliver point out [25]. They also argue that the way we use digital media and the internet is influenced by our skills, by competence and also by primary factors, such as self-control, media literacy, parent/child intervention, etc. [25]. All these are determined by social and cultural context that we experience [24]. Gui et al. point out that digital media, or technology, systematically shape our behaviour regardless of our features, and they also warn that in order to maintain digital well-being or quality of life in cyberspace, we need to introduce and form new aspects of digital skills [24].

As these authors explain, features that make digital technology or media useful (reliability, mobility, user-friendly approach and fast processing) can endanger our productivity and innovations but also our well-being as the stimuli patterns and patterns of individual reactions that are bound here are rather complex and specific. Combination of characteristics in this type of technology makes this cognitive and emotional dimension unprecedented and not neutral in relation to our opportunity to take part in a satisfactory communication practice. It seems to lead to a rapid and nonlinear use of information and communication. The authors warn that also those who are creative and have good social skills can constantly suffer from overcommunication [24], which, in our opinion, may be regarded as a risk that comes with our technically mediated internet-based interaction and communication in regards to quality of our life and digital well-being in the cyberspace of internet.

Self-control in using digital technology is simply not enough for us to cope with side effects of information overload, Gui et al. explain; complexity of modern media world wins this fight because self-control has always depended on moral values of the subject more than on any other competences. Therefore, we need to control digital stimuli and filter them so that they can serve our personal aims and well-being. We need to develop a new set of strategic resources, cognitive and metacognitive approaches and operational skills that influence our attention. These will lead to strategic approach in dealing with side effects of digital overcommunication. Such strategies should then serve as a prevention of stress caused by excessive information flow and also as a means to minimise wasting of time and attention on irrelevant activities in our everyday life [24].

Forming of these new aspects of digital skills is certainly a positive sign; however, we believe that quality life cannot entirely depend just on them, even though they, no doubt, contribute to digital well-being. If then we really aim for "good life", meaning more in axiological and ethical than psychological context, we must go even beyond.

We would argue that if we want to achieve "good life" in the internet cyberspace, based on high standards of morals and values, then the aspect of moral and humanly acceptable life, behaviour and actions, or good manners and positive principles, is relevant. We believe this is the condition for "ars vivendi". This is also the reason why we should use and improve elements that form it. We will be speaking about these elements in the following chapter.

5. "Ars vivendi" in the cyberspace of internet and education

As the specific living world of the cyberspace of internet shows, each person that is involved deals with aspects of life that are, to a great extent, given. We believe that we, creative beings, have a duty to "give our life a meaning and ensure it is coherent with our experience" [26] also in the context of our living world. In this world, our individual interests and social roles should create a coherent, even though not complete, life story [9].

Deuze suggests that this life should mean a piece of art, and it should be our life with ethic and aesthetic potential [10].[6] Our attitude is similar, and we think that in the context of using digital media, or with our life in the digital universe of

---

[6] We share his idea-he believes, following Z. Bauman, M. Foucault or F. Nietzsche, that our life can be a piece of art in which we all are actors, willingly or unwillingly, whether we realise it or not, and regardless of whether we enjoy it or not [10].

internet cyberspace, our life should follow the idea of "ars vivendi"-"good life" that is expressed by ethos-the way we live and deal with the others [27]. It is a way of life that becomes a prerequisite for us-intersubjectively communicative moral beings, a condition for every humane "learning to live" a humane life among other people.

Also Baeva points out that we, as human beings, still remain moral subjects even in the digital world of internet cyberspace, despite our virtual way or life that we lead here; we still keep our individual decision-making processes but, in addition, we also have new forms of freedom of moral choice. Values that media culture offers (freedom, personality orientation, pragmatism and others) become a new moral challenge for our behaviour, while ethical, axiological and value pluralism impose on us even bigger personal responsibility for our own moral and value choices [22].[7] What is important here, Deuze explains, is our ability to lead "our life in media" responsibly and safely in the internet cyberspace-in other words-make it safe, authentic and ethical. Deuze points out that this is our lifelong moral responsibility, even duty [10].

We agree with Deuze that we have individual responsibility to understand what we do in the internet cyberspace. However, there is a question that we need to ask here: who are we in this cyberspace? Deuze also points out that we should not lose oneself in the multitude of our own self-images and identities. We should use the internet, or digital media, in a way that both secures our independency and allows us to learn about ourselves at the same time. Deuze continues and explains that it can be difficult to find out who we really are in cyberspace, just as much as it is difficult to find out who we are in real life. This requires more emphasis on our own individual experience and understanding of the world. It is therefore necessary for us to contemplate our own life and existence [10].[8] This represents a path that leads to the required coherent self-understanding in our "ars vivendi" within the context of the internet cyberspace. In fact, we believe that it is indispensable.

According to Varanini, we need to prepare for life in the digital cyberspace [28]. The question is how a modern man can get prepared for life that should represent "good life" and "ars vivendi" in the digital universe of internet. If, for example, education is one of the social and cultural mechanisms that prepares individuals for life roles, then teaching us cope with everyday life becomes an actual task and challenge for education. Also Kačinová argues that the general goal of education, especially media education, is to prepare a student for life in the world of media [29] obviously also in the context of the internet cyberspace. The developing concept of media literacy, or digital literacy, meets this objective.[9]

Despite undergoing various changes in the past, it seems that understanding of digital literacy needs to be revised once again. Its concept needs to be refreshed. As far as we speak of digital well-being in the internet cyberspace, it is necessary to say that we are confronted with certain limits that (media) education should, or even must, deal with. We therefore believe that digital literacy should cover the new aspects of digital skills that are beneficial for our life in the context of the internet cyberspace and digital well-being. Education that provides relevant digital literacy applicable for our living world should include forming of the aspects that we mention above-skills, cognitive attitude and strategies that also Gui et al. speak about [24].

[7] Since simulated virtual reality is, by nature, reversible, temporary and never definitive and therefore always possible reality, it is necessary to be aware of our responsibility for our actions in cyberspace [22].

[8] Also Deuze points out that life in media inevitably brings multiple versions of "ME". So, who do we look for in media when we ask who we really are? Are all these versions that live in media equal? Are we able to distinguish between us and other individuals in media, or do we need to scrutinise bits and pieces? [10].

[9] Together with J. Suoranta and T. Vadén, we believe we can understand it as various processes of using digital information and communication technology to achieve the common good [30].

On the other hand, Deuze correctly points out that for quality life in the context of digital media, or good life with values and moral attitude, we need to avoid overestimating of media literacy for life in digital cyberspace and putting it above other skills. He continues, along with R. Rorty, that our culture employs instrumental rationality that favours knowing, expertise and professionalism, which prevent internal instability and increase immunity against romantic enthusiasm. This causes lack of inspiration, beauty and hope in our answers to challenges brought by "life in media" and therefore there is not enough inspiration, beauty and hope in the cyberspace of internet. According to Deuze, we should not be restricted by normative principles; in fact, we should use playful principles (tools and abilities) and also the virtue to feel astonished [10].

We believe that the above-mentioned attitude means that (media) education, which intends to prepare us for "good life" and corresponding values and moral attitudes in the internet cyberspace, should include also psychagogy.[10] This way it could prepare the ground for our self-reflexion and self-projection and also assist to improve our morality and self-control as something that is relevant for our "ars vivendi" in the internet cyberspace. We think that this education should lead us towards healthy scepticism in what we think is obvious and indisputable, and we should then be more active in our quest for the true, good and beautiful. However, it could also let us express our hopes, dreams and ambitions, and we should be free to wonder, appreciate and feel astonished. There could also be an opportunity for human modes such as slow speed, waiting, silence, boredom and emptiness[11], as well as for keeping one's distance or askesis in relation to digital media. We believe these are methods and elements that, when incorporated into the process of building digital literacy, can help us approach digital media and understand both ourselves and our living world in the internet cyberspace in a way that our "ars vivendi" requires.

6. Conclusion

It appears that we are reaching another milestone in our development and becoming "homo digitalis", moving to Cyberia, a cyberland [34]. As "homo digitalis," "we in fact become "homo cyberneticus", but also "homo medialis", "homo informaticus" and also "homo interneticus" or "homo smartphonus" dwelling in a specific world of the internet cyberspace. As this cyberspace is a virtual place, we can say that we become virtualised bio-socio-electronic subjects [22], and in this virtual space, we, human beings, think and act, behave certain way, project and express our ambitions, hopes, motives and goals. We simply live and take advantage of media technology. This way we are part of specific cyber experience, individual and collective at the same time. This comes with many positives, but also certain risks that in the long term may negatively influence its value.

If our experience in the communication-based living world of the internet is to bring us maximal value and enriching element, it has to offer the so-called digital well-being, which is one of the conditions and indicators of its quality. Along with this, we believe, in the axiological and ethical point of view, that life in cyberspace should also mean moral-value based "good life", which means appropriate values, behaviour and conduct, doing good-in other words, employ positive humane values and principles. This constitutes the "ethos", our style of life and actions visible for the others [27].

---

10 More on psychagogical dimension [31, 32].

11 These are "counter measures" of virtual, or media reality [33].

To conclude, who we are and how we live in the internet cyberspace is important. It is quite a significant issue as this way we intersubjectively define our life and values and share this definition with others. In fact, we inevitably take part in an intersubjective and collective process of "learning how to live". In this perspective, we are all confronted with the relevant "ars vivendi" with necessary coherent self-understanding and moral-value attitude. We believe that education that could really be beneficial should also include psychagogy. This could be a complementary aspect in building digital literacy and thus help us use digital media correctly and develop our self-understanding and understanding of the living world of the internet cyberspace by identifying values and morals that constitute our "ars vivendi".

Author details

Sabína Gálíková Tolnaiová and Slavomír Gálik*
Faculty of Mass Media Communication, University of Ss. Cyril and Methodius in Trnava, Trnava, Slovakia

*Address all correspondence to: s_galik@yahoo.com

References

[1] Holmes D. Communication Theory. Media, Technology and Society. London: Sage; 2010. p. 243

[2] Clark D. Characterizing Cyberspace: Past, Present and Future [online]. 2010. Available from: http://web.mit.edu/ecir/pdf/clark-cyberspace.pdf [Accessed: 15 December 2013]

[3] Groothuis D. Christian Scolarship and the Philosophical Analysis. JETS 41/4, December 1998. pp. 631-640. Available from: https://www.etsjets.org/files/JETS-PDFs/41/41-4/41-4-pp631-640-JETS.pdf [Accessed: 15 January 2020]

[4] Cappuro R. Beyond humanisms. Journal of New Frontiers in Spatial Concepts. 2012;**4**:1-12. Available from: http://ejournal.uvka.de/spatialconcepts/wp-content/uploads/2012/01/spatialconcepts_article_1362.pdf [Accessed: 15 January 2020]

[5] Brey P, Søraker JH. Philosophy of Computing and Information Technology. 2009. Available from: https://ethicsandtechnology.eu/wp-content/uploads/downloadable-content/Brey_Soraker_2009_Phil-IT-1.pdf [Accessed: 15 January 2020]

[6] Nanni C. Výchovná komunikácia: Modely, limity a to, čo ich presahuje [educational communication: Models, limits and what goes beyond them]. In: Kudláčová B, Rajský A, editors. Kontexty filozofie výchovy v novoveku a súčasnej perspektíve. Trnava: TU; 2014. pp. 83-93

[7] Macek J. Koncept rané kyberkultury [early cyberculture concept]. In: Média a realita. Brno: MU; 2003. pp. 35-61

[8] Cocking D. Plural selves and relational identity. Intimacy and privacy online. In: Van de Hoven J, Weckert J, editors. Information Technology and Moral Philosophy. Cambridge: University Press New York; 2009. p. 123-141

[9] Sprondel J, Breyer T, Wehrle M. CyberAnthropology – Being human on the internet [online]. 2011. pp. 1-77. Available from: http://www.hiig.de/wp-content/uploads/2012/04/CyberAnthropology-Paper.pdf [Accessed: 16 November 2014]

[10] Deuze M. Život v médiích. [Media life]. Karolinum: Praha; 2015. p. 268

[11] Turkle S. The Second Life. Computers and the Human Spirit. Cambridge: The MIT Press; 2005. p. 387

[12] Kendall L. Meaning and identity in "cyberspace": The performance of gender, class, and race online. Symbolic Interaction. 2011;**21**(2):129-153

[13] Matthews S. Identity and Information technology. In: Van den Hoven J, Weckert J, editors. Information Technology and Moral Philosophy. Cambridge: University Press New York; 2009. p. 142-160

[14] Višňovský E. Človek Ako Homo Agens. Ľudské Konanie Medzi mysľou a sociokultúrnym Kontextom. [the Man as Homo Agens. Human Action between Mind and Socio-Cultural Context]. Bratislava: Iris; 2009. p. 279

[15] Rankov P. Informačná spoločnosť – perspektívy, problémy, paradoxy [Information Society - Perspectives, Problems, Paradoxes]. Levice: LCA Publisher Group; 2006. p. 175

[16] Bauman Z. Lyon, D. Tekutý Dohled [Liquid Supervision]. Broken Book: Olomouc; 2013. p. 150

[17] Lévy P. Cyberculture. Minnesotta: University of Minnesota Press; 2001. p. 280

[18] Kováč T, Gyén M. Internet ako psychologický problém? [Internet as a psychological problem?] In: Vopálenský J, editor. Médiá na prahu tretieho tisícročia – človek v sieti mediálnej recepcie. Trnava: FMK UCM; 2003. p. 165-179

[19] Brožík V. O hodnotách a ľuďoch [About values and people]. Nitra: FF UKF; 2006. p. 177

[20] Kuzior A. Dekonštrukcia subjektu vo svete simulakier. [subject deconstruction in the world of simulacras]. In: Karul R, Porubjak M, editors. Realita a fikcia. Bratislava: SFZ pri SAV a KF FF UCM. 2009. pp. 246-251

[21] Bystřický J. Mediální realita [Media reality]. In: Magál S, Mistrík M, Solík M, editors. Masmediálna komunikácia a realita I. Trnava: FMK UCM; 2009. pp. 11-18

[22] Baeva VL. Existential and ethical values in an information era. Journal of Human Values. 2014;**20**(1):33-43

[23] McLuhan M. Understanding Media. London: Taylor & Francis Ltd; 2012. p. 400

[24] Gui M, Fasoli M, Carradore R. Digital well-being. Developing a new theoretical tool for media literacy research. Italian Journal of Sociology of Education. 2017;**9**(1):155-173

[25] Reinecke L, Oliver MB. Media use and well-being: Status quo and open questions. In: Reinecke L, Oliver MB, editors. The Routledge Handbook of Media Use and Well-Being: International Perspectives on Theory and Research on Positive Media Effects. New York: Routledge; 2016. p. 3-13

[26] Veugelers W. Introduction: Linking autonomy and humanity. In: Veugelers W, editor. Education and Humanism. Rotterdam: Sense Publisher; 2011. pp. 1-7

[27] Foucault M. Moc, Subject, Sexualita. Články a Rozhovory [Power, Subject, Sexuality. Articles and Interviews]. Bratislava: Kalligram; 2000. p. 233

[28] Varanini F. Human being in the digital world: Lessons from the past for future CIOs. In: Bongiorno G, et al, editors. CIOs and the Digital Transformation [Online]. 2018. Available from: <https://www.researchgate.net/publication/318831126_Human_Being_in_the_Digital_World_Lessons_from_the_Past_for_Future_CIOs> [Accessed: 20 February 2019]

[29] Kačinová V. Teória a prax mediálnej výchovy. Mediálna výchova ako súčasť všeobecného školského vzdelávania [Theory and practice of media education. Media education as a part of general school education]. Trnava: FMK UCM; 2015. p. 260

[30] Suoranta J, Vadén T. Wikiworld. Political Economy of Digital Literacy and the Promise of Participatory Media. University of Tampere, Paulo Freire Research Center [online] 2008. Available from: <https://wikiworld.files.wordpress.com/2008/03/suoranta_vaden_wikiworld.pdf> [Accessed: 20 September 2018]

[31] Gáliková Tolnaiová S. Problém výchovy na prahu 21. storočia (alebo o obrate k„psychagógii“ v súčasnej filozofii výchovy) [the Problem of Education at the Beginning of the 21st Century (or about Turning to "Psychagogy" in Contemporary Education Philosophy)]. Bratislava: Iris; 2007. p. 250

[32] Gáliková Tolnaiová S. Idea psychagógie v holistickej perspektíve [the Idea of Psychagogy in a Holistic Perspective]. Bratislava: Iris; 2014. p. 156

[33] Welsch W. Umelé rajské záhrady? Skúmanie Sveta elektronických médií a iných Svetov. [Artificial Garden? Exploring the World of Electronic Media and Other Worlds]. Bratislava: Soros Center for Contemporary Arts; 1995. p. 42

[34] Leary T. Chaos & Cyberculture. Oakland, California: Ronin Publishing, Inc.; 2014. p. 372

Section 2

New Technologies

Chapter 5

Cyberspace and Artificial Intelligence: The New Face of Cyber-Enhanced Hybrid Threats

Carlos Pedro Gonçalves

Abstract

While, until recently, cyber operations have constituted a specific subset of defense and security concerns, the synergization of cyberspace and artificial intelligence (AI), which are driving the Fourth Industrial Revolution, has raised the threat level of cyber operations, making them a centerpiece of what are called hybrid threats. The concept of hybrid threat is presently a key concern for the defense and security community; cyber-enabled and cyber-enhanced hybrid operations have been amplified in scope, frequency, speed, and threat level due to the synergies that come from the use of cyberspace and machine learning (ML)-based solutions. In the present work, we address the relevance of cyberspace-based operations and artificial intelligence for the implementation of hybrid operations and reflect on what this cyber dimension of hybrid operations implies for the concept of what constitutes a cyberweapon, the concept of hybrid human intelligence (hybrid HUMINT) and possible responses to the hybrid threat patterns.

Keywords: hybrid threats, cyber psychological operations, hybrid HUMINT, artificial intelligence, strategic studies, intelligence studies, data science

1. Introduction

The concepts of *hybrid threat* and *hybrid warfare* are, presently, key concepts within strategic studies[1] and intelligence studies[2], with a core relevance in the new defense and security context that was enabled by the twenty-first century's Fourth Industrial Revolution, driven by the synergization of *cyberspace* and *artificial intelligence* (AI), fueled by the accelerated and disruptive exponential expansion of machine learning (ML) [1–3]. Cyber operations, presently, constitute a key determinant component of *hybrid strategies* and *tactics* that configure the profile of *hybrid threats* and *hybrid warfare* [1]. *Hybrid strategies*, in the twenty-first century, involve the use of *Information and Communication Technology* (ICT) and AI tools to

[1] Strategic studies involve the study of strategy, crossing different disciplines, including military science, decision science, political science, and even systems science, and cognitive sciences.

[2] By *intelligence* we mean all the activities involved in the production of knowledge necessary to strategic and/or tactical decision. Intelligence studies are, then, the area of research that addresses all activities involved in such production of knowledge, including but not restricted to spying. The current chapter crosses, in permanent dialog, cyberspace studies, strategic studies, and intelligence studies.

combine conventional and unconventional operations, amplifying the impact of these operations [1–3].

In the current context of hybrid operations, there are, presently, three major dimensions of *hybrid strategic power*, understood as the ability to achieve one's strategic goals through *hybrid operations*, and these are:

- Network power
- AI power
- Cooperation power

The first type of power is enabled by social networks and the ability to use cyberspace for propaganda, disinformation, and viral campaigns in what constitutes a form of information-based warfare as well as for implementing cyberattacks that can disrupt different sectors as well as stealing (and possibly leaking) of critical data.

The second type of power involves the use of AI, in particular ML tools, as support tools for different cyber operations that may, in turn, support hybrid strategies. The range of AI applications can go from operations that take advantage of network power to cyber disruption of key infrastructures.

The third type of power is specific of today's defense and security environment, involving the cooperation of different state and non-state entities, the latter which include, for instance, organized criminal groups and terrorist groups that can cooperate with each other, supporting and enhancing each other's operations.

In the present work, we address the relevance of cyberspace-based operations and AI for the implementation of hybrid strategies and reflect on what this cyber dimension of hybrid operations implies for the concept of what constitutes a cyberweapon, as well as strategies that take advantage of the weaponization of cyberspace. We also address the concept of human intelligence (HUMINT) operations and their role in hybrid operations, in particular, how HUMINT was used in the past to support hybrid operations and can play a key role in the present; this leads us to the conceptualization of *hybrid HUMINT*.

In Section 2, we review main concepts linked to hybrid operations and address the strategic profile of hybrid operations in its different dimensions.

In Section 3, we focus on *cyber psychological operations (cyops)* as a major part of *hybrid strategies* and address how the use of AI and ML in *cyops* can be employed for the operationalization of hybrid strategies, targeted at weaponizing social networks, showing that AI constitutes a central driver of the future of these operations and allowing us to produce an assessment of the near future of hybrid threats, including a new face of *cyber terrorism*.

In Section 4, we address another dimension of hybrid operations and hybrid threats which is the role of HUMINT and the concept of *hybrid agent*, reviewing how HUMINT was used in the past for the implementation of hybrid operations and how it can be used in the present as a nexus for the successful implementation of these operations. In Section 5, we conclude with a final reflection on the role of cyberspace and the need for an extended concept of *hybrid resilience* as a way to face hybrid threats.

2. Cyberspace and the strategic profile of hybrid operations

Hybrid operations can be defined as the use of military and nonmilitary means to achieve one's strategic goals [1–3]. This means that rather than open battle, one may use intelligence activities, subterfuge, and subversion in order to gain an advantage over the adversary.

Hybrid operations find a deep tradition in strategic thinking that can be traced back to the classics of strategic studies, in particular, to two of the main military classics of Ancient China [1, 4]: *Sun Tzu's Art of War* and *T'ai Kung's Six Secret Teachings*. These two works also inspired Japanese classical thinking about unconventional warfare and espionage and the use of specialized operatives that also implemented what can be considered today as hybrid operations. Operations with strategic and tactical dimensions were recorded during the transition from the Warring States period to the Edo period in different works. Of these different works, the *Sandai Hidensho* stands out, which consists of the scrolls that include the *Bansenshukai* [5], the *Shoninki* [6] and the *Shinobi Hiden* [7], these are three classical works on spying and on how to conduct subversive, covert, and unorthodox warfare, which also recognize the influence of *Sun Tzu's Art of War* and *T'ai Kung's Six Secret Teachings* [5–7].

While there is a deep tradition for hybrid operations in both Chinese and Japanese classics on warfare and spying, it is also important to stress that a thinking that is convergent with the Asian classics is also found in European Philosophical thinking about strategy and war, in particular in Machiavelli's *The Art of War* [8], which also addresses what are considered today as operations that fall within the scope of hybrid operations in books six and seven of this work.

In *T'ai Kung's Six Secret Teachings*, hybrid operations included corrupting key officials, using diplomacy as a weapon, compromising a kingdom's economy, alienating the ruler from the people, spreading rumors, and using propaganda and what is known today as psychological warfare [4]; similar operations are also described in the main Japanese classic on the art of spying, the *Bansenshukai* [5].

In what regards hybrid operations, the strategic action is not, thus, restricted to the battlefield but rather includes acting on the economic, financial, social, and political levels as a way to avoid open warfare or to weaken the adversary so that if open warfare does take place, one can easily win over that adversary [4, 5].

Specifically military hybrid operations are covered in the *T'ai Kung's Six Secret Teachings*, particularly in the *Dragon Secret Teaching*, in the section corresponding to the *unorthodox army*, and in the section addressing the *civil offensive*, in the *Martial Secret Teaching* [4], which is convergent with both the Japanese classics [5–7] and the European thinking [8].

One lesson that comes out of these classics of strategic studies and intelligence studies is the need for good governance and public policies as a way to guard against hybrid operations [4, 5], a point to which we will return in the last section of the present chapter. Disrupting governance goes to the key role of hybrid operations in classical strategy and intelligence thinking to undermine a country's governance and to make the people turn against the policymakers.

This is a point that is recovered in today's defense and security environment, present in different countries' military thinking. On the Russian side, as stressed by Chekinov and Bogdanov [9], two Russian Defense specialists, information technologies make the new face of warfare to be dominated by information and psychological warfare.

The central driving forces behind the twenty-first century's hybrid operations are *cyber psychological operations* (*cyops*), where *information superiority* plays a key role [1, 3, 9]. As stressed in [9], the new face of conflict is such that nonmilitary actions and measures are employed with ICTs used in order to target *all public institutions in a target country*. While this illustrates the Russian perspective on the twenty-first century conflict [1], we get a similar standpoint from Treverton [3], who is a former Chairman of the US National Intelligence Council. Treverton identified, in the pattern of hybrid operations, typical information *cyop*-based warfare tactics, using propaganda, fake news, strategic leaks, funding of organizations and supporting political parties, organizing protest movements (taking advantage of social networks), using cyber tools for espionage, attack and manipulation, economic leverage, use of proxies and unacknowledged war and supporting paramilitary organizations.

While deeply rooted in the past thinking of strategic studies and in past military practice, the above references [1–3, 9] show that the renewal of the concept of hybrid operations and the relevance of this concept in the twenty-first century strategic thinking and doctrine come from the fact that these operations now have an effectiveness amplified by the use of cyberspace, which is a determinant factor in the change of the profile of the defense and security threats coming from hybrid operations; more properly, as it is addressed in [1], the twenty-first century hybrid operations can be implemented by both state and non-state agents, and this implies a major shift in strategic power, where individuals and groups, which may not be state-sponsored, can use cyberspace and even AI-based systems to implement hybrid operations that can have significant impact on a given country's governance [1, 2].

This adds a new dimension to hybrid threats, making the profile more complex from a defense and security standpoint, in the sense that we can have three types of hybrid operations' profiles:

- **Type 1:** state-sponsored operations implemented by a specific country or countries: these are implemented by countries and involve the human and technical resources of that country's Armed Forces.

- **Type 2:** non-state-sponsored operations: these are implemented by non-state agents and groups, not supported financially, politically, and logistically by any state.

- **Type 3:** state-sponsored operations implemented by non-state agents: the use of hackers and techno-mercenarism, the political, financial, and logistic support to non-state agents and groups opens up the way for the implementation of joint operations that involve non-state agents and different countries (with an added level of plausible deniability for countries).

These three types of operations are key for the characterization of hybrid operations. The *type 1 operation profile* has always been an integral part of strategic thinking and doctrine regarding unorthodox strategies and tactics and the way in which one may win one's goals without using conventional military forces, an approach that is considered in high regard within the context of the Chinese classics [4] and that is recovered also in the Japanese context of the employment of specialized operatives called *shinobi no mono* that were used as spies and specialists in covert operations, subversion, information warfare, and what are considered in the *T'ai Kung's Six Secret Teachings* as unorthodox ways [5–7]. Currently, however, cyberspace has amplified the effectiveness potential of these unorthodox ways, making hybrid operations a core dimension of military doctrine and the twenty-first century conflict, a point argued extensively in [1–3, 9].

However, the state-sponsored hybrid operations, implemented by a country's armed forces, intelligence agencies, or even specific *cyber warfare units* and, possibly, *hybrid warfare units*, are just part of the three types of hybrid operation profiles.

The *type 2 operation profile* is characteristic of a change in the strategic power dynamics due to cyberspace and availability of AI systems and is specific of the new defense and security framework of hybrid threats, namely, small groups, or even a sufficiently knowledgeable individual, with sufficiently sophisticated hacking skills, can perform hybrid operations, taking advantage of cyberattacks and AI tools and target a country's governance, significantly disrupting that country with the same effectiveness as any state-sponsored attack. The threat ecosystem is, thus, no longer just one of the countries fighting each other but also of countries' governments and infrastructures being threatened by non-state agents that can implement hybrid operations as disruptive as any *type 1 operation*.

The key to the issue is the fact that cyber tools and even AI systems are freely available and the exponential trend linked to AI and ML and the increased usage of connected devices and smart government solutions open up the way for an exponential increase in the ability and opportunities to attack a country's governance with a low budget, this increases the disruptive potential of *type 2 operation profile*, which can be evaluated in terms of the increasingly low cost availability of means (including freely available bots and open-source malicious code dispersal), the increased dispersal of targets (due to the exponential trend associated with the Internet of Things (IoT)), and the ability to use cyberspace, including the dark web, to connect with like-minded individuals that are willing to support viral campaigns against specific targets.

Given the Fourth Industrial Revolution's foreseeable trend, the *type 2 operation profile* is typically a profile that involves most operations in cyberspace, given the high impact and low cost of these operations. The ability of sufficiently motivated individuals and groups, sometimes involved with criminal organizations, to successfully implement a hybrid operation with the same level of impact as a state-sponsored campaign is a point that only recently has been addressed in the literature on hybrid threats [1, 3, 10, 11], a point raised in [10]. This is a gap in that literature since there may be an underestimation of rapidly emergent threats. The main problem lies in the fact that hybrid operations can be implemented with significantly less investment, especially if their main component is cyberspace-based, and this can be considered as low-cost warfare or, as stated in [11], *war on the cheap*.

The Fourth Industrial Revolution has opened up the ability for weaker opponents, both state and non-state, to effectively engage opponents with stronger military forces, decreasing the comparative advantage of these stronger opponents. The network power and AI power allow for a non-state agents to launch a hybrid campaign on a targeted country from anywhere in the world, such that one may have difficulty in ascribing a given physical/national territory to the attacker and single out that attacker's country for a targeted conventional military response. A sufficiently sophisticated group can remain anonymous and even be transnational in the composition of its members, transitioning the defense problem from the traditional military dimension to a more complex response nexus of defense, intelligence, and law enforcement.

While *type 2 operation profile* is now being recognized as an increasing threat [1, 10, 11], with the tendency to increase in disruptive ability in the years to come, the *type 3 operation profile* has the potential for the most damage in that it involves the joint cooperation between state and non-state agents. This last operations' profile takes advantage of the cooperation power; cooperation in hybrid operations can take the form of cooperation between different non-state agents, including terrorist groups and different criminal organizations; between different countries; and between state and non-state agents.

The cooperation between state and non-state agents may become a main source of state-sponsored hybrid threats [10, 11]; rather than engaging in large-scale state-on-state conflict, different states can support non-state agents or act in a timing that is confluent with the actions of non-state agents, enhancing the ability of non-state agents to produce a large disruption on a targeted country's governance.

We are reaching a strategic context where both state and non-state agents can engage any given country by means of cyber operations, sabotage, espionage, and subversion [11], a point that also circles back to *Sun Tzu's Art of War* [4]. Hybrid operations, whatever their profile, allow an opponent or opponents to produce an imbalance of power, acting on the target's weaknesses, without engaging in conventional direct conflict and, possibly, even hiding their identities in the process.

The imbalance of power is linked, in Sun Tzu's thinking, to the concept of power as the ability to exercise one's authority and deliberative autonomy toward effective action; in this case, hybrid operations directly target a state's power by undermining its governance.

These operations can, in particular, take advantage of:

- Internal challenges to a state's governance by certain groups that wish to undermine a state's authority

- Failure of a state in adapting to society's concerns and its people's problems, being unable to respond to disruptions to the state's finances, economic problems, environmental problems, and social and political problems

- A view in a country's society that a regime has lost its legitimacy to rule

The above three targeted state-level weaknesses match what Margolis in [11] identified, respectively, as sources of three types of crises:

- ***Crises of authority*** that result from a state's inability to enforce its rule, not being able to control all of its territory, or becoming unable to enforce all its laws

- ***Crises of resilience*** that result from a state's inability to adapt to different disruptions

- ***Crises of legitimacy*** that result from society's view that *a regime has lost its right to rule because it is wrong or unjust*

These three types of crises can occur in a given country and be triggered by hybrid operations or amplified by well-timed hybrid operations, in particular, those that use information and cyberspace as a weapon.

Connecting the three profiles and crises, in **Figure 1**, we synthesize, in scheme, the links between the profiles of hybrid operations and the three types of crises identified by Margolis [11].

The *type 1* and *type 2 operations* are confluent with each other in the cooperation involved in *type 3 operations*. It is important to notice that, in some cases, a *type 1* or a *type 2 operation* can lead to a *type 3 operation*, and that the timing of a *type 1* with a *type 2 operation* can lead to a *type 3 operation* due to synchronized hybrid tactical actions. The three types of operations can all target the three weaknesses, authority, resilience, and legitimacy, amplifying state instability and leading a country into a crisis situation that may, in the limit, produce the fall of a government.

Now, regarding the means and vulnerabilities, it is important to stress that the Fourth Industrial Revolution also opens up the way for cyber-physical attacks, including attacks using drones and drone swarms, as well as cyberattacks on automated systems and cyber-physical systems; all these are dimensions of the wider cyber-enhanced synergy of conventional and unconventional operations that constitute the strategic and tactical ground for hybrid operations and that may predictably characterize the new level of hybrid operations in years to come.

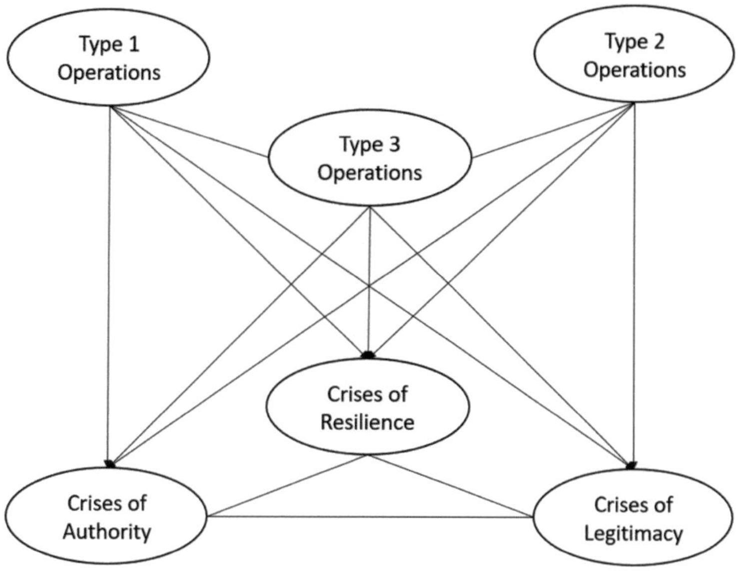

Figure 1.
Hybrid operations profiles and crisis profiles.

The other side, which we are already seeing today, is situated in the virtual space but still able to severely affect countries' governance; as stated above, this is the weaponization of cyberspace, using social networks and AI for hybrid operations. In this case, the actions are situated only in the virtual space, but they can have severe social, (geo) political, and economic consequences.

Platforms, in particular social networks, the manipulation of contents, and the use of AI, ML, and data science to manipulate people's behavior, online and offline, are a major component in these operations, and it is the subject of the next section.

3. Cyber psychological operations and hybrid threats

The strategic level of hybrid operations involves the definition of the main objectives for hybrid operations, the targets, and possible collaboration networks. The choice of resources and ways to combine them to operationalize the hybrid strategy depends upon the strategic deliberation. On the other hand, the means also condition the set of available tactics that may allow one to operationalize a given strategy.

The strategic power of hybrid operations in allowing for a state or non-state agents to achieve their strategic objectives has increased due to the resources available that allow for high yield with low investment; these resources are linked to the network power and AI power, defined at the beginning of the present chapter.

In what regards hybrid operations, the network power and AI power cannot presently be considered separately, since it is precisely the synergy of cyberspace and AI, in particular through ML, that determine the present strategic and tactical momentum of hybrid operations and that allow one to anticipate the future of hybrid threats. We now address one of the major components of hybrid operations, namely, information warfare and psychological operations using cyberspace.

Psychological operations (*psyops*) involve the use of different means and tactics in order to influence the behavior of target audiences. While, traditionally, *psyops* were employed by countries and constitute an integrating part of military doctrine, the expansion of cyberspace has led to the possibility of groups that are not part of any country's official military branch to implement these operations. An example of this is ISIS' online propaganda as well as hacktivist groups such as anonymous online activities.

The use of cyberspace and hacking, including the defacement of a country's websites, the online dispersal of sensitive and/or compromising data through social media platforms, the possibility of using the *dark web* for the disclosure of sensitive data that can then be made public in the *surface web*, and the use of social media for propaganda and recruitment, for the denouncement of different causes, and for the manipulation of citizen journalism as a way to publish both true and fake news as well as to disperse other fake contents (including images, audio, and videos), all these are examples of ways in which *psyops* can be implemented using cyberspace, so, at present, *psychological operations* are an integral component of hybrid warfare in what constitute *cyber psychological operations* or *cyops* for short.

As stated previously, in the present chapter, *cyops* are a major part of hybrid operations, and *cyber psychological tactics* involved in *cyops* typically include [3]:

- *Propaganda* (in particular, dispersed online through social media)
- *Fake contents* (in particular, *fake news*)
- *Online dispersal of sensitive data* (*leaks*)

Each of these tactics takes advantage of network power, AI power, and cooperation power. There are three drivers that have amplified the effectiveness of the above tactics:

- The increased dispersal of connected devices, including *smartphones* and *tablets* that allow an easy and frequent access to the Internet
- Search engines and online services that adapt to each user's interaction pattern
- The growing use of social media over traditional media

Added to this infrastructural accessibility to these devices is the high frequency use of these devices and sometimes addictive component associated with this use, an addictive component usually linked to social networks.

A specific pattern of usage favors the dispersal of sensitive data, news, and general contents in social media: the fact that the online reading of social media contents usually does not involve a high level of reflection but rather engages the users in a way that is meant to be appealing and to be shared quickly with as most people as possible, users seldom read or reflect deeply on the contents that they are sharing, usually skimming through them and sharing the most appealing ones.

This is a pattern that is particularly useful for dispersal of contents that are presented in the form of scandals, sensitive information that was not known, conspiracies' denouncements, and so on. This point leaves a marker in data on fake content dispersal as shown in a study on the differential diffusion of verified (true) and false rumors on Twitter from 2006 to 2017, published in [12]. In the study, *politics* and *urban legends* stand out as the two categories with the highest frequency in rumor cascades.

The study concluded that rumors about politics, urban legends, and science spread to the most people, while politics and urban legends exhibited more intense viral patterns [12].

The study found a significant difference in the spread of fake contents vis-à-vis true contents, namely, true contents are *rarely diffused to more than 1000 people,* while

DOI: http://dx.doi.org/10.5772/intechopen.88648

the top 1% of fake rumor cascades are *routinely diffused between 1000 and 100,000 people* [12]. The authors' results showed that fake contents reached more people at every depth of a cascade, which the authors defined as *instances of a rumor spreading pattern that exhibit an unbroken retweet chain with a common, singular origin*.

The result that fake contents *reached more people at every depth of a cascade* means that more people retweeted fake contents than true ones, a spread that was amplified by a viral dynamics. The authors found that fake contents did not just spread through broadcast dynamics but, instead, through peer-to-peer diffusion with viral branching.

Another relevant point, for hybrid operations, was that fake political contents traveled deeper and are more broadly reaching more people and exhibiting a stronger viral pattern than any other categories and diffusing deeper more quickly. This dynamics is not however due to users who spread fake contents having a greater number of followers; the study found exactly the opposite with a high statistical significance. In inferential terms, users who spread fake contents tend to have fewer followers, to follow fewer people, to be less active on Twitter, are verified less often, and have been on Twitter for less time. However, fake contents were 70% more likely to be retweeted than true contents with a *p-value* of 0.0 in Wald chi-square test.

The fact that user connectedness and network structure did not seem to play a relevant role in *fake content* dispersal made the authors seek other explanations for the differences in *fake content* versus *true content* dispersal. The authors reported that fake contents usually inspired greater number of replies exhibiting surprise and disgust. The authors' hypothesis is that novelty may be a key factor in *false rumor dispersal*.

However, there is a relevant point to take into account when looking at the study's results, which can be expressed by the following extreme example: an account with no followers and not following anyone can still get a high number of retweets and exposure on a content if it uses *hashtags* on hot topics and builds its tweet in a specific way that increases the probability of it being retweeted.

Moving beyond this specific study and considering social networks in general, working with the conceptual basis of strategic studies, we are led to introduce the concept of *tactical accounts*, defined as accounts that are created for tactical purposes in the support of a *cyop* strategy; these accounts can be managed by a single individual or staffs, and its operations can involve the use of *bots* that automatically generate contents with certain specifications, mostly aimed at making the contents viral in the spread.

The viral content design along with multiple accounts operated by *bots* are major tools for a *tactical account system manager*, that is, any operative can use multiple tactical accounts simultaneously to create a fake content dispersal so that it can gain momentum and become viral.

In general, fake contents can spread on hot topics by the use of *hashtags* or other means of dispersal, which diminishes the connectivity need for any single *tactical account*'s effective impact. Furthermore, from a *cyops*' standpoint, it is easier to *fly under the radar* by managing multiple newly created fake accounts that can even be managed by a single agent, who may then use these accounts to disperse fake contents incorporating *hashtags* on political issues and composing the messages so that they have an appealing emotive content, making it more likely for people to select them.

Returning to the study [12], the fact that the authors did not find strong evidence that algorithms were biased toward spreading of fake contents but rather that fake contents were being dispersed by people is favorable to the point of the way in which the message is built as the key factor in getting a fake content to gain traction. This point is echoed in [13] where it is argued that the belief in fake contents is driven by emotional responses amplified by macro social, political, and cultural trends.

Social media are particularly sensitive to the careful crafting of the message to fit viral conditions, in the sense that these media are managed by platform-based businesses, optimized for quick spread of information to reach target audiences and mass dispersal; in this sense, they are aimed by design at viral dynamics and addictive usage patterns that increase the interaction time with the platform and create value for these businesses.

The technology is thus an enabler of viral dynamics and, in that way, facilitates fake contents' dispersal by the way in which these contents are produced, in terms of the message that they contain, the emotional responses which they are aimed to evoke, and their timing and their management of conditions of dispersal (for instance, the use of *hashtags* on trending topics in Twitter); all this contributes to the increased likelihood that fake rather than true carefully crafted and reflection demanding content become viral.

Hybrid tactics can take advantage of multiple (*fake*) *tactical accounts* and use methods of automation of content generation, with possible applications of data science, in order to generate the content presentation that may be the most effective in getting people to adhere to and, thus, share. By working with data on viral tweets, ML algorithms may be trained in predicting the structure of a content that may make it more probable to become viral and use this to help a *cyops* operative design the message in order to make it more viral and then use *tactical accounts* to disperse it. Message contents, including *hashtags*, *emoticons*, and *gifs*, are useful tools in manipulating the message content to better fit a target audience [14].

Another way to manipulate viral content dispersal is cyberattacks aimed at compromising search engines and recommendation engines in order to disperse fake content that fits the goals of an intended *cyop*.

Search engine poisoning or even a more sophisticated *search engineering* has been applied in the past by *black hats* to spread malware and fake contents [15, 16].

There are various methods employed in this last context that can be highly effective for hybrid operations: the first is content injection in websites, online forums, and social media in the form of spam posts that can also point to specific websites used within a *cyop*; this is a task that can be automated.

A second level is the creation of networks of websites and social media accounts that spread alternate media messages and that reinforce *echo chambers* for specific content that can, thus, become viral, taking advantage of a concerted social media campaign that divulges these accounts.

The sharing of these accounts can, in turn, become viral and link to different alternate media websites that can be used for *cyops* and manipulate a user's web search and interaction with different content platforms. If, in the interaction, with any search engine and content platform, there is a powerful algorithmic adaptation to each user's pattern, then, any user, influenced by viral content, will have a tendency to be fed back the content that the *cyop* is aimed at. In this way, by strategically using viral dynamics, a *cyop* can manipulate a vast amount of users and engineer massive *echo chambers* where massive amounts of users get personalized content that fits the *cyop* in question.

In this case, the hacker or hackers do not need to compromise the AI systems that manage a social media platform; rather, they are *hacking people's behaviors* and are taking advantage of the effectiveness of the platform's own AI systems in adapting content to user interaction profile. Since the way a user interacts with a platform leads to a specific response on the part of the platform for automatic user personalization, each user gets his/her own experience; however, the commonality of usage patterns allows for collectives of users with common tastes to receive similar or confluent viral contents.

Creating and financing tactical networks of social media accounts amplify this *hybrid strategy*, as long as the platform adapts very quickly to a user's profile facilitating the *echo chamber engineering* needed for the *cyop* to be successful. Similar tactics can be employed on any type of social network. However, of the different online media platforms, Facebook seems to stand out in terms of effectiveness of fake news dispersal, with a higher frequency of cases of visits to fake news websites occurring near a Facebook visit, as reported in [17].

Besides content injection in *blogs*, *forums*, and *social media*, another way for search poisoning involves content injection in compromised websites, as well as search redirection. Search redirection attacks employ sites that have been compromised to be used in a search redirection operation and whose owners usually do not suspect that their website has been compromised [16]. These *source infections* in turn redirect to traffic brokers that redirect traffic to specific destinations that fit the hackers' main goal [16]. Currently, ML algorithms are being trained against redirection as a defense against it [18]; however, ML algorithms and data science can also be employed to manipulate content, including written text, pictures, and even videos. In the foreseeable future, a higher ability of *deep fake videos* to fool people may greatly enhance the impact of fake content dispersal.

While disinformation and propaganda, through online fake content and propaganda dispersal operations, have become highly impactful in terms of their strategic and tactical value [3, 17], there is another level of *cyops* that may be implemented by any state or non-state agent that can have a strong impact on society. This is exemplified by the *Blue Whale Challenge*, which is an example of the power of what can be considered a *gamification attack*.

Gamification attacks use the Internet to introduce a game which leads the players through a series of challenges down a path where those players are led to either self-harm or even murder. In the case of the *Blue Whale Challenge*, the players were led to self-harm. The game involved a series of life-threatening tasks given to players by a curator, and each player had to fulfill these tasks which ended with the suicide of the player [19]. In a certain sense, this constitutes a cyberspace-enabled form of murder, by leading a person to commit suicide. The *Blue Whale Challenge*'s curators can be treated as a new breed of serial killers that use the Internet for psychological and physical torture, eventually leading their victims to kill themselves as the endgame of the tasks that they give their victims.

If we replace the final task of *suicide* with a final task where the player has to *murder* someone else or even a number of people, perhaps even in exchange for the player's own life (an *either kill yourself or commit murder* option), then, the *Blue Whale Challenge* becomes the first example of designing a web game that can lead not only people to suicide but also to murder on a scale and intensity that can be comparable to those of standard terrorist networks.

One should stress that this is a form of *cyops* that can easily be engineered by someone not affiliated to any terrorist group. A single person can take advantage of the power of cyberspace and of social networks to create such *challenges*; furthermore, even if the individual is caught and arrested, the game can go on independently of the individual, where anyone can become a *curator*. The game itself becomes the terror referent and the platform for terror practices.

This breaks with any traditional approach to engaging and handling terrorist organizations, since a *terror game* can be played by anyone, without any political goal, without any political affiliation, and with no end other than the exercise of violence. These new serial killers that become curators of these games can be caught and imprisoned, but the game can go on with different iterations. There is a form of digital autonomy and continuation of a *terror* game as a collective dynamics that is sustained by its players, but that goes on despite the catching of particular curator players, as long as it is available for playing; the game can even come back with new variations and remain, and even if it has no players, it can be played again at any time.

This is not a *terror network* that one can address with traditional tactics; it is a *terror game*, and the *Blue Whale Challenge* is just the first example of this. The game becomes the *referent* for any players, who may never have physically met. Systemically, the game becomes a *dispositional driver* for a typological order of cyber-enabled terrorist practices. Another point is that, potentially, such *terror games* can be sustained by non-humans, that is, by AI systems, and even if all human curators were caught and arrested, *bots* could take over and play the same role as a human curator (the player that abuses the other players). In this sense, a single individual, using AI systems, can create *terror games*, sustained by an "army" of *cyop bots* that will be difficult to stop. A new breed of the twenty-first-century serial killers can become a source of new cyber-enabled terrorism that uses *gamification* as a way to resiliently murder on a global scale, with an impact on par with that of major standard terrorist organizations.

The reason why *bots* can be used here as *cyber psychological weapons*, in such *games* as the *Blue Whale Challenge*, is linked to the algorithmic basis of these games' approach; in particular, the behavior of curators can be algorithmically replicated by *cyop bots*. Indeed, the process involves using social networks to search for young people who fit specific profiles, which can include being depressed or showing addictive behavior. The list of tasks includes dynamics that introduce sleep deprivation, listening to psychedelic music, watching videos with disturbing contents sent by the curator, and inflicting wounds on one's body, among other tasks [20]. The tasks follow a prescribed set of steps that lead the victim into a disturbed mental state and susceptible to the influence of the curator, the victim is a target of a form of *cyop* that falls within a pattern that can easily be turned into an algorithm.

The *gamification* of *cyops* in terror operations is in its infancy; however, the tools available to it are amplified by the IoT, mobile devices, and platform usage. In the *Blue Whale Challenge*, we see a new tactics based on platform weaponization, that is, the use of platform-based businesses to compromise its users and eventually lead to their deaths (in the case of the *Blue Whale Challenge*) or even to the killing of others (if instead of suicide the player is led to kill others).

Empowered by *cyop* bots, a small number of individuals, or even one individual, can create a game that may go on independently of them; the game can persist as a dynamics that continues to be played in the platform, which functions as a replicator for the deviant and predatory behavioral patterns needed for the *terror game to go on*. Having been played once, the dynamics that characterize the game can always come back; in this sense, the platform works as a way for the digital continuation of the terror game.

This is very different from the case of a terrorist network that has a hierarchical structure and that has cells and individuals that play different roles within an organization.

A *terror game* is just a set of behavioral patterns, with algorithmic components, that can be replicated like a form of social virus which goes on as long as there are players. There is no stable hierarchy and no cells and no individuals that can be targeted which may harm the game, because the game has a virtual fluid existence that can be perpetuated as a dynamics to be retrieved any time, any place.

The *terror game* is characteristic of a side of platforms, especially social networking platforms that make them highly weaponizable, namely, platforms are means for the exercise of biopower in the sense of Foucault [21], a point that is convergent with the issues addressed in [22].

Platforms can function as means for the exercise of control, reward, and punishment and of manipulation of its users' desires, fears, and sources of inclusion and exclusion, integration and segregation, connection and isolation, and friendship and bullying.

By increasingly sharing one's life in platforms and by using integrated systems, in particular IoT devices, the new stage of the Internet revolution is such that any heavy user of these systems can be *datafied*, profiled, and manipulated by hacked devices (including hacked AI systems) and manipulated by predators that use fake accounts and their victims' profiles to launch directed *cyops* that can, in the end, as was the case with the *Blue Whale Challenge*, lead to a person's death.

According to data, reported in [20], Instagram ranks higher in posts than the Russian VK social network (which was where the game spread initially) and Twitter. On Twitter, the large majority number of posts related to the *Blue Whale Challenge* was identified by the authors as coming from *smartphones* with the Android OS, which shows how mobile devices are useful in feeding *terror gamification* operations.

Another pattern revealed in these authors' research is a key common factor in online *cyop* campaigns. In particular, many accounts talking about the *Blue Whale Challenge* were new accounts with not many followers; this shows again the possible use of *tactical accounts*. This is a basic necessary tactical choice for predators operating online, who will want to hide their identity; furthermore, in order to gain online traction on a *cyop*, the use of multiple *tactical accounts* is a necessary step. Thus, just as in *state agents*, *non-state agents*, including *cyber-enabled serial killers*, may tend to use multiple *tactical accounts* in online platforms when addressing their targets.

The use of challenges like the *Blue Whale Challenge* and the *Momo Challenge* directly targets a large amount of victims and constitutes a security and law enforcement problem [23].

Returning to *cyops*, whatever their profile, these are currently about using cyberspace and ML for *hacking people*'s behaviors. In this sense, while a *cyop* against a given country may take advantage of resilience, authority, or legitimacy vulnerabilities, the increasing use of the platform-based technologies, managed by ML algorithms, where each user's data is exposed and available for exploitation, leads to another level of vulnerability which is the ability to use citizens' own data and behavioral patterns against them or to manipulate citizens into patterns of behavior that interest a given state or non-state agent.

The fact that *cyops* have certain components that are algorithmizable implies that one can program bots as *cyop* weapons that function as a form of new computer virus, a behaviorally conditioning content-based virus that is aimed at hacking people's behaviors, delivered through platforms for both mass exposure and personalization. The current trend of using algorithms for decision-making and in everyday life, integrated in platforms and that feed on each user's data and adapting the service and contents to each user's profile, makes AI weapons, employed in *cyops*, increasingly effective tools.

While the *cyops* that were discussed above include the creation and manipulation of contents to produce responses and manipulate people's behaviors, the impact of these contents can become even more amplified if the dispersal of these contents is timed with the leak of true contents. In this case, people tend to believe the fake content that is consistent with the true content. The leak of true content can initiate a fake content campaign, where the true content provides the context for the fake contents that will be used in the fake content campaign.

In this case, *leak platforms*, like *WikiLeaks*, can be used by hackers, whistleblowers, as well as other agents (state and non-state agents) to disperse true content and provide the timing for initiating fake content campaigns. However, besides *leak platforms*, there is another level of hybrid operations which also increases the threat of these types of operations for any country; this is the new *hybrid human intelligence/counterintelligence (CI)* context, in which the concept of a new field agent is a key factor.

4. Hybrid HUMINT

The concept of human intelligence involves a twofold dimension: to gather information from human sources that are not HUMINT operatives and to gather information from HUMINT operatives. In terms of operations, HUMINT involves [24]:

- The clandestine acquisition of relevant data
- The overt collection of relevant information by people overseas
- The debriefing of foreign nationals and citizens who travel abroad
- Official contacts with foreign governments

While budget restrictions, the cyberspace expansion, and the development of data science have fueled the interest in signals intelligence (SIGINT) and open-source intelligence (OSINT) and led to some divestment in HUMINT, considered, for instance, more expensive in terms of time and resources involved than OSINT, an approach that considers an opposition of HUMINT *vs* OSINT and HUMINT *vs* SIGINT is the wrong way to look at things from an *intelligence/counterintelligence* effectiveness standpoint, within the new defense and security context, characterized by the critical threat of hybrid operations.

In fact, one can robustly argue, from a technical and technological standpoint, that HUMINT is a major centerpiece driver of hybrid operations, a nexus around which SIGINT and OSINT can be leveraged, with the new agents on the ground being able to both gather strategic and tactical information, implement (cyber) subversive maneuvers, steal data, and even compromise critical systems of any organization.

In the new context of hybrid operations, a new breed of HUMINT operative is not only a spy but also a hacker and a hybrid operations specialist that can infiltrate an organization and bring it down from the inside. We call this the *hybrid agent*.

From a *counterintelligence* standpoint, the new dimensions of the threat of covert human agents need to be critically addressed. The first thing to stress is that the threat level is very high for any state; on the other hand, the operational advantage of the new breed of HUMINT operative is also very high, so that, from an *intelligence/CI* standpoint, states need to invest in both these new *hybrid agents* and to find countermeasures for them.

To fully realize the implications and measures of the concept of a *hybrid agent*, which is the main point of this section, we need to first address some conceptual dimensions from intelligence studies and strategic studies, since while the tools of the *hybrid agent* have changed and the profile and impact is new, there was an old case of a form of *spy* that fit this profile of *hybrid agent* which was employed in Japan's Warring States period (*Sengoku Jidai*) and later in the Edo period. This old *hybrid agent* fits a similar profile and role that the new *hybrid agent* may come to fit in the years to come.

During the *Sengoku Jidai*, spies were mainly employed from the Samurai and Ashigaru classes, but progressively, especially in the Iga and Koka provinces, spying was systematized, developed, and integrated in a body of knowledge and skills that were taught to warriors, a body of knowledge that was built on top of the warrior normal training.

Different regions and Samurai clans also had their trained spies. It is important to stress at this point that there coexisted two types of spies in Japan: warriors who were employed as spies but that were not trained spies and warriors who, besides their normal martial training, were trained as spies. Another division that arose was between the trained spies that were a part of a Daimyo's army and thus served the Daimyo and the spies for hire, mercenary spies.

Due to their skills, Iga and Koka spies became mercenary spies, that is, spies for hire that also operated based on alliances of these regions with different groups. It is important to consider what constitutes the body of knowledge that the Japanese incorporated in what they considered to be the *art of spying*, called *shinobi no jutsu* or *ninjutsu*, erroneously addressed in popular culture as a martial art, as assassination, and/or as warriors that opposed the samurai, all incorrect misconceptions [25].

The fact that traditional scrolls on this body of knowledge are hard to track down, being, in many instances, in private collections, in some way contributed to the misconception to be perpetuated and has produced a gap in the literature on intelligence which usually cites *Sun Tzu's Art of War* but overlooks, in the study of the history of intelligence, the deep development of the theory, strategies, and tactics of intelligence that is present in the traditional texts on *shinobi no jutsu*, which, as a relevant point in dispelling the misconception, never cover any kind of hand-to-hand fighting techniques [5–7, 25, 26].

Recently, thanks to the efforts of the historian Antony Cummins and Yoshie Minami, the major texts are now translated into modern English, and Cummins has tracked down scrolls beyond the main known texts and translated them to English, making them available to the wider audience.

These texts allow people, researching in intelligence studies, to find new references that deepen *Sun Tzu's Art of War*'s last chapter. These works, in particular [5, 26], develop in great detail a full body of knowledge in what the Japanese considered the art of spying and operationalize *Sun Tzu's Art of War*'s last chapter into a detail that provides an insight into the history of intelligence.

The relevant point is that these works on *shinobi no jutsu* introduce a profile of an operative, the *shinobi no mono*, which is largely a *hybrid warfare specialist*, and also constitute some of the few examples of classical works that are only devoted to intelligence, that is, these are some of the few examples of *Classical Intelligence Manuals* that form a compendium of the main strategies and tactics of intelligence in Japan's Warring States and Edo periods, some of which hold, in terms of their main patterns and principles, for any period and place.

If one analyzes these different classical Japanese works on what are today called intelligence studies [5–7, 25], one finds that, in Japan, the *art of spying* (*shinobi no jutsu*) included, among other specialized knowledge, a core of areas of expertise that, under close scrutiny, are generalizable to other countries and historical periods [27], and these areas include:

- Military strategy and tactics
- Scouting
- Infiltration and tactical disruption
- Unconventional warfare, including deep knowledge of subversive maneuvers and psychological operations
- Deep knowledge of counterintelligence

Infiltration came in two ways [5]:

- *Yojutsu*: which involved infiltrating the enemy in plain sight, that is, using long-term undercover agents
- *In jutsu*: which involved stealing in, hiding from the enemy

In jutsu was largely employed during the *Sengoku Jidai* and already included the conventional and unconventional, where the trained agents infiltrated the enemy ranks, usually at night and used fire and unconventional tactics to disrupt the enemy in a way that allowed for a well-timed conventional open attack to ensue. These were the precursors of battlefield hybrid operations and are documented in detail in the *Bansenshukai* [5].

Yo jutsu usually needed someone with some level of scholarship, namely, from the Samurai class. This was the long-term undercover operative, which not only gathered information just like any HUMINT specialist but also employed subversive maneuvers, disinformation, counterintelligence, and political manipulation. One finds an example of this in the *Bansenshukai* [5], which reportedly is an Iga manual, but that may also contain a synthesis of Iga and Koka knowledge, where it is stated that an agent needed to obtain and keep copies of the marks and seals of the lords of various castles so that these could be used to forge letters in order to incriminate a target for conspiracy. Using agents to frame key people sow discord among the enemy's ranks and even for assassination (in particular, through poisoning).

Undercover operatives (using *yo jutsu*) were employed for disrupting the enemy's intelligence and decision-making process (disinformation), making false charges, spreading rumors (the dispersal of fake contents was already present in this period), and sowing domestic conflicts, discord, and doubts among the enemy's vassalage, as well as for setting fires or causing confusion among the enemy's castle in order for an open strike to occur. These are documented in the *Bansenshukai* [5] and are all parts of hybrid operations. Indeed, the play between the conventional and unconventional which is a key characteristic of *shinobi no jutsu* is strongly convergent with the two the major Chinese classics of strategy: *Sun Tzu's Art of War* and *T'ai Kung's Six Secret Teachings* [4], the latter which is considered in [5, 7] a Chinese reference on what the Japanese called *shinobi no jutsu*.

The Japanese knew of both works, and they are referenced in the different Japanese classical texts on *shinobi no jutsu* [5–7, 25, 26]. In particular, Sun Tzu's five types of agents were employed and elaborated upon in terms of intelligence strategies and tactics in the context of the Japanese classics, and these five types are [4]:

- Local spies (employing of locals to gather information)
- Internal spies (employing people who hold government positions)
- Double agents (employing the enemy's agents)
- Expendable spies (employed to spread disinformation outside the state; in the Japanese case, they used these in conjunction with the highly trained undercover operatives, which, due to their training, were not considered expendable but were, rather, high-valued assets that were used for what are today considered the core of hybrid warfare: disinformation, psyops, fake content and rumor spreading, sowing discord and unconventional strategies and tactics, besides spying per se)
- Living spies (who returned with their reports)

In [26] these five types of spies are explicitly addressed with an in-depth analysis on the strategies and tactics that are involved in their usage.

To these five types, one can add another type, which holds a key value for the new hybrid operations' context: the unwitting agent, who is supplying information for the enemy but is unaware of this fact. One can already find this type of agent in some passages of the *Bansenshukai* [5].

Now, taking into account this historical context, let us consider what we called the twenty-first century *hybrid agent*, which, in terms of operational profile, is used in the same manner as the *shinobi no mono*, namely, we have an expert in strategy and tactics that can be infiltrated in an organization (*yo jutsu*) and who will be used to both gather critical information (the standard classical HUMINT aspect) and find the main vulnerabilities of the target organization and, if given the activation order, is capable of disrupting the organization from the inside using cyberattacks, compromising key employees, releasing compromising data, and/or launching a fake content campaign against the organization.

Mirroring the setting of fire and the compromising of the intelligence cycle used in Medieval Japan, we now have the possibility of a long-term undercover operative to physically install malware and cyberweapons and hack critical systems to corrupt key data and disrupt the organization's normal functioning.

Business, banking, healthcare, and government are particularly vulnerable sectors that can be hacked in this way. The businesses' use of ML-supported OSINT can be compromised by malware aimed at attacking the ML infrastructure and thus corrupt strategic decisions, business secrets, and strategically sensitive data which can be stolen to undermine a target country's business (state-sponsored corporate espionage) either by using these data for gaining a negotiation leverage, an R&D and competitive advantage or, simply, to disclose it, bringing losses to these businesses.

Companies and banking that employ platforms, in the new 4.0 paradigm, can have their platforms compromised by a *hybrid agent*, undermining the stakeholder confidence.

If there are corruption practices or any key figures in key business, banking and/or political sectors fall prey to entrapment (even digital entrapment); then, this can be used to disrupt a country's business, banking, and even political sectors, even to turn the people against these sectors in well-orchestrated hybrid campaigns that use social networks to amplify the disruption effect. The principles behind this *economic and political warfare*, which is a key dimension of hybrid strategies, are expanded in detail in *T'ai Kung's Six Secret Teachings*.

Our main point is that the new *hybrid agent* is a key player in making this type of hybrid operations effective. The reason for this is that while remote hacking, SIGINT, OSINT, and even *cyber intelligence* (CYBERINT) can be effective, the *hybrid operative* is more disruptive and may greatly enhance SIGINT, OSINT, and CYBERINT; there are a few reasons for this.

Hacking an organization becomes exponentially more effective if it is done by someone who is undercover inside the organization, and this person can have direct access to an organization's critical systems and compromise them by physically installing malware. Social engineering also becomes easier and more effective if combined with direct personal interaction with human targets that can be hacked. Hacking colleagues' smartphones, for instance, and other IoT devices can lead to a new form of *unwitting agent*: the person who takes his/her devices everywhere, devices that can be accessed by the *hybrid operative* and used to record audio, video, geographical data, and other personal data. This means that conversations can be recorded, video can be recorded, and even personal data can be gathered and used to compromise a target individual.

With increasing sensorization of organizations, a successful *hybrid operative* can turn the organization's *sensorization* systems into his/her own listening devices. Furthermore, standard HUMINT can be combined with OSINT and SIGINT, where the *hybrid operative* can directly interact with a human target, hacking the target's devices, employing social engineering tactics, and then combining the cyber intrusion with fake social network accounts, managed by a remote team that may follow the target on such places as Facebook, Twitter, Instagram, and so on, further interacting with this human target, using the social media, private chat systems, and even video chat sessions with remote support team operatives, in order to manipulate the target and find the target's weaknesses, gaining the target's confidence and possibly compromising the target or using that target as an (unwitting) source of information.

The trained *hybrid operative* must then be:

- An expert in *cyops*
- An expert in *hybrid operations*
- A hacker with strong skills in social engineering

From a CI standpoint this is a major threat on two fronts:

- On the state-sponsored front: the *hybrid operative* is a key nexus for combining synergistically HUMINT, OSINT, SIGINT, Social Network Intelligence (SOCINT), CYBERINT, and *cyops*, taking all this to a new level which can seriously disrupt a country's key public and private organizations.
- On the non-state-sponsored front: a very skillful hacker team or even an individual hacker, with strong social engineering skills, who have physically infiltrated a target and are supported by *bots* that automate the fake content dispersal, can, with very low cost, produce the same effect as a trained state-sponsored team.

The second front is a major problem, since it opens up the way for new *hybrid warfare mercenarism*; just as the Iga and Koka *shinobi no mono* were employed as mercenaries, it also opens up the way for non-state-sponsored hybrid attacks from individuals or groups that have a cause or even just a grudge against a target, individuals, and groups who are skilled hackers that can perform similar operations as a *hybrid agent*.

In this sense, there can be three operational profiles for *hybrid agents* which mirror the three operational profiles for hybrid threats addressed in Section 2:

- Type 1 hybrid agent: an agent that belongs to a given state's intelligence agency and that is operating covertly
- Type 2 hybrid agent: an agent not linked to any intelligence agency but highly skilled in hacking and social engineering that is not operating on behalf of any state but is either a lone wolf or operating on behalf of some non-state group
- Type 3 hybrid agent: an agent not linked to any intelligence agency but that performs hybrid operations for hire

The three types of agents may coexist and constitute a major threat for countries' national security and defense; on the other hand, one may also recognize that, while constituting a threat, any state may take advantage of these three types of agents in its own operations, with particular relevance to types 1 and 3 as well as the relevance of the tactical openings provided by the actions of type 2 agents. There is a fluid border between the three types, where agents can change their profile along the course of their activities.

The question that can be raised is: *what responses need to be implemented in terms of CI to deal with the twenty-first century hybrid agent?* The answer is somewhat complex, in the sense that the threat landscape is changing with the exponential technological revolution that greatly enhances the disruptive power of the new hybrid HUMINT, which can synergistically combine traditional with high-tech methods to become one of the most disruptive forces in the new defense and security context, but, given an identification of major targets, in particular economic and financial targets (that may become key parts of economic, financial and political warfare), there are specific responses that need to come into play with some urgency. This forms part of our final reflection on the whole chapter and is integrated in the next section, which concludes the chapter.

5. A final reflection and possible responses

Throughout the chapter we laid out the profile as well as the current and foreseeable evolution of hybrid operations and hybrid threats (Section 2). We also addressed the issue of weaponization of cyberspace, the use of AI and data science, and the threat patterns of *cyber psychological operations* in the context of hybrid operations (Section 3), and, in Section 4, we introduced the concept of *hybrid agent*, evaluating its overall pattern of activity and threat to countries' defense and security.

Some major points need to be highlighted, when dealing with *hybrid threats*, namely:

- Operations on the virtual space can have physical consequences, even in the cases where the operation does not directly disrupt physical systems.

- Related to the previous point, behavioral hacking is a major component of *cyops* and can take advantage of the impact of fake contents, propaganda, disinformation, as well as strategic leaks of critical data, in order to affect people's behaviors.

- *Gamification* and implementation of viral online *challenges* can support *terror games* that may gain a form of digital continuity, such that the game can be recovered anytime, even after the arrest of key individuals and groups, being perpetuated independently of what happens to the initiators of these *terror games*, and can be kept going by *bots* as well as by people willing to play the game, taking advantage of the dynamics between AIs and social media.

- A new form of operative, the *hybrid agent*, leads to an amplification of the synergy between HUMINT, SIGINT, and OSINT with HUMINT playing the nexus role, in which an undercover agent takes advantage of the physical presence on any given organization and employs classical HUMINT strategies and tactics along with hacking and *cyops* to enable and enhance the disruptive potential of well-orchestrated *hybrid campaigns*.

These are some major points that were addressed in detail in the previous sections. Now, as part of a final reflection, the question may be raised: *what to do about all this?*

From the work developed throughout the sections, one thing becomes clear: there is an urgent need for the strategic integration in key state and private organizations, including the defense and security community, of a concept of *hybrid resilience*, of which *cyber resilience* is just an aspect. In this sense, in what regards CYBERINT [28], its focus needs to address the profile of cyber-threats and *cyop* profiles associated with hybrid strategies, in the sense that tactical dynamics of cyberattacks may obey to the pattern needed for a given hybrid strategy, and it needs to cooperate with HUMINT/CI in order to find countermeasures against *hybrid HUMINT* operatives.

The concept of *hybrid resilience* as the ability to resist and recover from *hybrid campaigns* should be a major component of countries' national defense and security strategies.

Now, secondly, organizations should have training and a *hybrid defense and CI division* or at least subcontract specialized people in this area, covering both *cyber defense* and *cyber resilience* as well as *hybrid defense* and *resilience*.

Faced with the threat of economic, financial, and (geo) political hybrid warfare, any country's major business and financial targets should have specialized training programs and people involved in *hybrid defense strategies* and *hybrid resilience*, including CI-based defense against possible disruption from what may become the new disruptive face of HUMINT: the *hybrid HUMINT*.

It is not enough to secure the technical side of cybersecurity, and one needs to address the social and human aspect of cyber intrusion, in which people's behavior can be turned against them, including the behaviors and vulnerabilities that come from incorrect social network usage.

Campaigns in the standard media against fake contents need to be addressed, as well as large-scale educational programs that should start in schools, educating civil society on the correct usage of cyberspace, on both the positive and negative, on how people can protect themselves against cyberbullying and hybrid campaigns, and on how people should read and reflect on the contents that they read and share.

While these are some of the major changes needed to be implemented for any country's successful *national hybrid defense strategy*, there is a main point of *hybrid resilience* that was already identified in the old Chinese and Japanese classics, in particular in *Tai Kung's Six Secret Teachings* [4] and in the *Bansenshukai* [5]: without good governance there is always a fundamental vulnerability to hybrid strategies.

The three major crisis profiles that were addressed in [29] and recovered in Section 2 come out of bad governance that is unable to face crises that affect its country's people (*resilience problems*), that is totalitarian and oppressive and that enforces its rule by force or has alienated a large part of its people due to rising inequalities and widespread political, business, and financial corruption (*legitimacy problems*), or that is unable to manage its territory (*authority problems*). All these three problems open up any country to *hybrid threats* and reduce a country's *hybrid resilience*.

DOI: http://dx.doi.org/10.5772/intechopen.88648

Author details

Carlos Pedro Gonçalves
Institute of Social and Political Sciences, University of Lisbon, Lisbon, Portugal

*Address all correspondence to: cgoncalves@iscsp.ulisboa.pt

References

[1] Atkinson C. Hybrid warfare and societal resilience: Implications for democratic governance. Information and Security: An International Journal. 2018;**39**:63-76. DOI: 10.11610/isij.3906

[2] Nikolic N. Connecting conflict concepts: Hybrid warfare and warden's rings. Information and Security: An International Journal. 2018;**41**:21-34. DOI: 10.11610/isij.4102

[3] Treverton GF. The intelligence challenges of hybrid threats: Focus on cyber and virtual realm. Sweden. Center for Asymmetric Threat Studies. 2018. 36p. ISBN: 978-91-86137-75-5. Available from: http://fhs.diva-portal.org/smash/get/diva2:1250560/FULLTEXT01.pdf [Accessed: 04 June 2019]

[4] Sawyer RD. The Seven Military Classics of Ancient China. United States: Basic Books; 2007. 568 p. ISBN:0-8133-1228-0

[5] Cummins A, Minami Y. The Book of Ninja: The First Complete Translation of the Bansenshukai, Japan's Premier Ninja Manual. London: Watkins Publishing; 2013. 512 p. ISBN: 978-1-78028-493-4

[6] Cummins A, Minami Y. True Path of the Ninja: The Definitive Translation of the Shoninki. Singapore: Tuttle; 2011. 191 p. ISBN: 978-4-8053-1114-1

[7] Cummins A, Minami Y. The Secret Traditions of the Shinobi: Hattori Hanzo's Shinobi Hiden and Other Ninja Scrolls. Berkeley: Blue Snake Books; 2012. 194 p. ISBN: 978-1-58394-435-6

[8] Machiavelli N. Art of War. Chicago: The University of Chicago Press; 2003. 262 p. ISBN: 0-226-50040-3

[9] Chekinov SG, Bogdanov SA. The nature and content of a new-generation war. Military Thought: A Russian Journal of Military Theory and Strategy. 2013;**4**:12-23

[10] Normak M. How States Use Non-State Actors: A Modus Operandi for Covert State Subversion and Malign Networks. The European Centre of Excellence for Countering Hybrid Threats. 2019. Available from: https://www.hybridcoe.fi/wp-content/uploads/2019/04/HybridCoE_SA_Non-state-Actors_RGB_NEW.pdf [Accessed: 24 April 2019]

[11] Raugh DL. Is the hybrid threat a true threat? Journal of Strategic Security. 2016;**9**:1-13. DOI: 10.5038/1944-0472.9.2.1507

[12] Vosoughi S, Roy D, Aral S. The spread of true and false news online. Science. 2018;**359**:1146-1151. DOI: 10.1126/science.aap9559

[13] Nisbet EC, Kamenchuk O. The psychology of state-sponsored disinformation campaigns and implications for public diplomacy. The Hague Journal of Diplomacy. 2019;**14**(1-2):65-82. DOI: 10.1163/1871191X-11411019

[14] Nemr C, Gangware W. Weapons of Mass Distraction: Foreign State-Sponsored Disinformation in the Digital Age. 2019. Park Advisors. Available from: https://www.state.gov/documents/organization/290985.pdf [Accessed: 28 April 2019]

[15] Howard F, Onur K. Poisoned Search Results: How Hackers have Automated Search Engine Poisoning Attacks to Distribute Malware. SophosLabs Technical Paper. 2010. Available from: https://www.sophos.com/en-us/medialibrary/PDFs/technical%20papers/sophosseoinsights.pdf [Accessed: 28 April 2019]

[16] Leontiadis N, Moore T, Christin N. A nearly four-year longitudinal study of

search-engine poisoning. In: Proceedings of the 2014 ACM SIGSAC Conference on Computer and Communications Security. New York: ACM; 2014. pp. 930-941. DOI: 10.1145/2660267.2660332

[17] Guess A, Nyhan B, Reifler J. Selective Exposure to Misinformation: Evidence from the Consumption of Fake News During the 2016 U.S. Presidential Campaign. European Research Council. 2016. Available from: https://www.dartmouth.edu/~nyhan/fake-news-2016.pdf [Accessed: 06 May 2019]

[18] Hans K, Ahuja L, Muttoo SK. Performance evaluation of neural network training algorithms in redirection spam detection. In: Panigrahi B, Hoda M, Sharma V, Goel S, editors. Nature Inspired Computing. Advances in Intelligent Systems and Computing. Vol. 652. Singapore: Springer; 2018. pp. 177-183. DOI: 10.1007/978-981-10-6747-1_20

[19] Lupariello F, Curti SM, Coppo E, Racalbuto SS, Di Vella G. Self-harm risk among adolescents and the phenomenon of the "Blue Whale challenge": Case series and review of the literature. Journal of Forensic Sciences. 2019;**64**(2):638-642. DOI: 10.1111/1556-4029.13880

[20] Khattar A, Dabas K, Gupta K, Chopra S, Kumaraguru P. White or Blue, the Whale gets its Vengeance: A Social Media Analysis of the Blue Whale Challenge. 2018. Available from: https://arxiv.org/pdf/1801.05588.pdf [Accessed: 10 May 2019]

[21] Foucault M. Microphysics of Power (Microfísica do Poder). Edições Graal: Brazil; 1979. 295 p. ISBN: 9788570380746

[22] O'Neil C. Weapons of Math Destruction—How Big Data Increases Inequality and Threatens Democracy. New York: Crown; 2016. 259 p. ISBN:978-0-553-41881-1

[23] Błasik P. Social networks and personal security. Scientific Journal of Bielsko-Biala School of Finance and Law. 2018;**22**(3):13-16. DOI: 10.19192/2543-411X-3

[24] INTelligence: Human Intelligence. 2010. Available from: https://www.cia.gov/news-information/featured-story-archive/2010-featured-story-archive/intelligence-human-intelligence.html [Accessed: 29 May 2019]

[25] Cummins A. In Search of the Ninja: The Historical Truth of Ninjutsu. The UK: The History Press; 2012. 239 p. ISBN: 978-0-7524-8093-0

[26] Cummins A, Minami Y. Iga and Koka Ninja Skills: The Secret Shinobi Scrolls of Chikamatsu Shigenori, Including a Commentary on Sun Tzu's 'Use of Spies' in The Art of War. UK: The History Press; 2014. 189 p. ISBN: 978-0-7509-5664-2

[27] Hughes-Wilson J. On Intelligence. UK: Constable; 2016. 528 p. ISBN: 978-1-47211-354-2

[28] Clark RM, Oleson PC. Cyber intelligence. Intelligencer: Journal of U.S. Intelligence Studies. 2018;**24**(3): 11-23 Available from: https://www.afio.com/publications/CLARK_OLESON_Pages_from_Vol24_No3_AFIO_INTEL_Winter_2018-19_FINAL.pdf [Accessed: 04 June 2019]

[29] Margolis JE. Estimating state instability. Studies in Intelligence. 2012;**56**(1):13-24. Available from: https://www.cia.gov/library/center-for-the-study-of-intelligence/csi-publications/csi-studies/studies/vol.-56-no.-1/pdfs-vol-56.-no.-1/Estimating%20State%20Instability%20-Extracts-Mar12-20Apr12.pdf [Accessed: 24 April 2019]

Chapter 6

Combined Deep Learning and Traditional NLP Approaches for Fire Burst Detection Based on Twitter Posts

Konstantinos-George Thanos, Andrianna Polydouri, Antonios Danelakis, Dimitris Kyriazanos and Stelios C.A. Thomopoulos

Abstract

The current chapter introduces a procedure that aims at determining regions that are on fire, based on Twitter posts, as soon as possible. The proposed scheme utilizes a deep learning approach for analyzing the text of Twitter posts announcing fire bursts. Deep learning is becoming very popular within different text applications involving text generalization, text summarization, and extracting text information. A deep learning network is to be trained so as to distinguish valid Twitter fire-announcing posts from junk posts. Next, the posts labeled as valid by the network have undergone traditional NLP-based information extraction where the initial unstructured text is converted into a structured one, from which potential location and timestamp of the incident for further exploitation are derived. Analytic processing is then implemented in order to output aggregated reports which are used to finally detect potential geographical areas that are probably threatened by fire. So far, the part that has been implemented is the traditional NLP-based and has already derived promising results under real-world conditions' testing. The deep learning enrichment is to be implemented and expected to build upon the performance of the existing architecture and further improve it.

Keywords: deep learning, NLP procedure, fire burst detection, twitter posts, valid posts

1. Introduction

Due to their cost and easy access, social media and Twitter, among them, are widely used as sources of news and means of information spreading. Among others, fire bursts are such breaking news that can be initially made known through Twitter posts.

Mega fires often result in significant environmental destructions, major damages on infrastructures, and economic loss. Most importantly, they put at stake the lives, not only of the civilians but also of the forest fire personnel. Thus, technologies that facilitate early fire detection are important for reducing fires and their negative effects.

Our approach proposes the combination of a deep learning architecture along with a more traditional natural language processing (NLP) one. The deep learning component of the system is responsible for filtering out the fake from the valid fire-related posts, so that only posts containing true fire-related information are retained. For this part of the system, we refer to current state-of-the-art systems for detecting fake news and adopt the one that suits the needs of our problem best. Once the fake posts are filtered out, each valid post is afterward fed into the NLP-based subsystem. By converting the unstructured, raw text into a structured one, the NLP-based subsystem is able to extract information, such as the geographical area of the fire reported in the post. In order to draw final conclusions about the possible fire sources, aggregation statistics over the posts containing similar fire-related information are computed, and probability values for each potential fire source are given as output.

The rest of the chapter is organized as follows: Section 2 describes and analyzes the deep learning-based architecture to be utilized for detecting valid Twitter posts regarding fire bursts. Section 3 illustrates the typical NLP-based architecture for extracting meaningful information from the unstructured text of a valid Twitter post. Section 4 presents the overall scheme and its final output. Finally, before the conclusions, Section 5 highlights the results of the up-to-date validated part of the overall proposed scheme.

2. Deep learning-based architecture for false Twitter post detection

2.1 Introduction

Social media are low-cost and easy-to-access means of information sharing and, thus, nowadays are widely used as source of news and information. However, getting informed from social media is not always safe, as posts expressing fake news (i.e., news containing false information) are exponentially widespread, simultaneous to the boosting development of online social networks. In fact, fake news tends to outperform the valid ones in the near future [1].

In case of fire burst news, deciding whether a Twitter post is fake or not can be proven of crucial twofold importance. On the one hand, the required time and money for purposeless activation of the firefighting mechanisms are saved. On the other hand, timely confrontation of mega fires is facilitated. This will, in turn, make it less likely for human lives, the environment, and infrastructures to be jeopardized.

Thus, before extracting the crucial fire burst information at a later NLP-based stage, a preprocessing step, deciding whether a Twitter post that declares a fire burst is fake or not, is necessary. To this end, a deep learning-based architecture is to be implemented. The purpose of this architecture is to filter out the posts that will be characterized as "fake" and provide the sequential NLP procedure only with the "valid" posts.

2.2 Candidate deep learning architectures

In this subsection, the candidate state-of-the-art deep learning architectures for the detection of fake posts are described. The purpose of the subsection is to illustrate the most recent and modern approaches that have appeared from 2017 onward and have been examined. The input data, used by all architectures, is the text provided by social media posts, especially focused on Twitter. The output is the decision whether the input text corresponds to a "valid" or "fake" post.

The 3HAN architecture [2] utilizes a three-level hierarchical attention network. Each of the three levels corresponds to words, sentences, and headline analysis. The three-edge analysis results in the construction of a news vector which represents the input post. The latter vector is used for classifying the reliability of the post.

The architecture presented in [3], namely, ConvNet, uses a convolutional layer to capture the dependency between the text and its metadata. For the case of the metadata, a standard max pooling and a bidirectional Long Short-Term Memory (LSTM) auto-encoder layer follow. For the case of the text, only a max-pooling layer is implemented. Finally, the max-pooled text representations are concatenated with the metadata representation from the bidirectional LSTM. The merged concatenations are fed to a fully connected layer with a softmax activation function. This generates the final prediction.

The work in [4] presents the FakeNewsTracker architecture. This is a deep learning architecture which is divided into two sub-schemes. The first sub-scheme uses an LSTM deep network [5] in order for the system to be trained on the post representation context. The second sub-scheme utilizes a recursive neural network (RNN) in order to be trained on the context of social engagements. The output features of the aforementioned sub-schemes are fused together to perform a binary classification procedure which labels the input news as "fake" or "valid."

The DeClarE architecture [6] is based on bidirectional LSTMs in order to result in a credibility score related to the input post. The scheme also considers post source and claims information, which is processed within the bidirectional LSTM dense layers. The concatenated output is also processed by two dense layers and a softmax layer before the prediction of the credibility score.

The work in [7] introduces a hybrid architecture approach which combines an LSTM and a convolutional neural network (CNN) model. Throughout this chapter, the aforementioned architecture will be called Hybrid LSTM-CNN. The LSTM was adopted for the sequence classification of the data. The 1D CNN was added immediately after the word embedding layer of the LSTM model. A max-pooling layer is also recruited to reduce dimensionality of the input layer, thus avoiding training over-fitting of the training data. This also helps in reducing the resources for the training of the model.

The FakeDetector architecture [8] relates post creators to posts and subjects. It contains a Hybrid Feature Vector Unit (HFLU) which extracts the feature vector based on a specific input. The feature vector is fed to the gated diffusive unit (GDU) model for effective relationship modeling among news articles, creators, and subjects. Formally, the GDU model accepts multiple inputs from different sources simultaneously. The GDU applies softmax operation on the output vector before assigning a credibility label. For a more explicit sight on deep learning architectures, the reader is referred to [9].

2.3 Procedural requirements

Before deciding which architecture fits best in our specific case, the direct requirements of the overall procedure should be recorded.

To begin with, detecting fake posts in real time is an essential requirement of the process. Rapid decision whether a fire-bursting declaration post is fake or not leads to fast implementation of the NLP procedure (as described in Section 3). The latter, in turn, facilitates the timely detection of the geographical areas threatened by fire as soon as possible which helps toward the prevention of the majority of negative effects caused by mega fires. Therefore, the proposed architecture of [3] is not suitable for our use case, as it is not implemented in a fully automated manner.

Fake news detection accuracy is very important. High detection accuracy guarantees that the great majority of the posts that fed to be processed in the sequential NLP phase (see Section 3) express sincere fire burst claims. Thus, the final resulting fire-threatened geographical areas are much more likely to be actually threatened. Furthermore, the aforementioned accuracy needs to have been achieved in publicly available datasets and benchmarks. This windows the performance of the architecture much more reliable than others, tested on proprietary datasets. To this end, the FakeNewsTracker architecture [4] is not suitable for our use case, as it is tested on a proprietary dataset.

Last but not least, the architecture needs to be domain invariant. In other words, it needs to be generally applicable to any domain, other than the one(s) used for conducting training and testing procedures. More precisely, the accuracy of a system, detecting fake post that deal with fire burst, should not be significantly altered in the case of post that deal with any other domain (politics, sports, etc.). This makes the system architecture much more flexible and adoptable. From the remaining architectures analyzed in this section, only DeClarE [6] and the Hybrid LSTM-CNN [7] claim to be domain invariant. DeClarE has been tested on PolitiFact dataset [10] achieving accuracy 67.32%, while the Hybrid LSTM-CNN has been tested on PHEME dataset [11, 12] achieving 82.00% accuracy. Both datasets are publicly available. PolitiFact is a respected fact-checking website releasing a list of sites manually investigated and labeled. It mostly contains posts of political content. PHEME is also another EU-funded project whose results include collecting and annotating rumor tweets which are associated with nine different breaking news contents. Therefore, PHEME is a richer dataset with a wider variety of themes that makes the Hybrid LSTM-CNN system architecture [7] it has been tested on more suitable for our use case.

The procedural requirements for the fake post detection scheme with respect to the architectures analyzed in Section 2 are summarized in **Table 1**.

2.4 Implementation architecture

Based on the aforementioned requirements, the baseline of the architecture selected to be implemented follows the Hybrid LSTM-CNN architecture [7]. The overall architecture is illustrated in **Figure 1**. The *input layer* consists of Twitter posts which are, in fact, unstructured raw texts. A *word embedding layer* follows, within which the input text is parsed and is divided into a series of words and, consequently, into a series of sentences.

Architectures	Procedural requirements			
	Real time	Accuracy	Public dataset	Domain invariance
3HAN [2]	✓	✓	✓	×
ConvNet [3]	×	×	✓	×
FakeNewsTracker [4]	×	✓	×	×
DeClarE [6]	i	✓	✓	✓
Hybrid LSTM-CNN [7]	✓	✓	✓	✓
FakeDetector [8]	✓	✓	✓	×

Table 1.
Procedural requirements for fake post detection part.

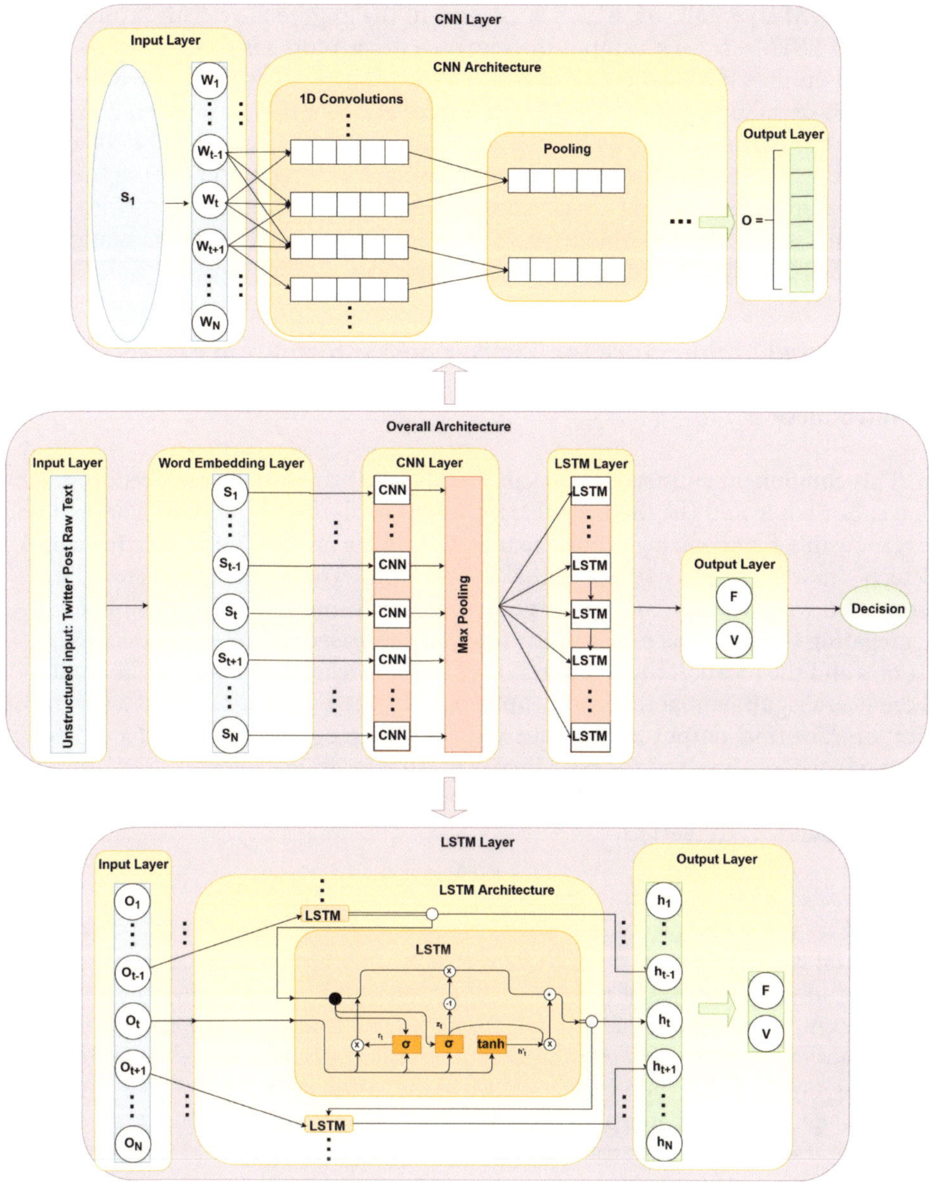

Figure 1.
Suggested fake post detection architecture.

Each sentence is then consumed by the *CNN layer* of the architecture which is made up of a set of 1D CNNs based on the work presented in [13]. The CNNs of this layer are structured as illustrated in the upper part of **Figure 1**. The 1D convolutions, taking place within the CNNs (as defined by Eq. (10) of the Appendix), operate on sliding windows of the words of the sentence. Before outputting the outcome of the layer, max pooling is performed to reduce dimensionality and avoid over-fitting of the training data. This also helps toward reducing computational complexity of the training process. The output of each CNN is a fixed length vector, acting as a digital signature of the corresponding sentence and describing the nature of the sentence. Thus, a set of such description vectors (descriptors) are fed forward for further process.

The *LSTM layer* follows, which is the core of the architecture. This layer consists of a set of LSTMs. It uses as input the sentence descriptors resulting from the CNN layer and outputs the final decision vector indicating whether the claim of the post is fake (F) or valid (V). Each LSTM of the layer is structured as presented in the lower part of **Figure 1**. LSTMs are chosen because they are proven to be robust for representing a series of data, such as the one we are dealing with here (i.e., series of words or sentences), as they are capable of capturing their internal temporal dependencies [9]. The LSTM layer is very interesting in terms of mathematics. For more information the reader is referred to Appendix.

3. NLP-based architecture for Twitter post information extraction

3.1 Introduction

This component consists of two sub-modules: (a) the fire incident report detection sub-module and (b) the fire incident report analytic sub-module. The first one is responsible for acquiring reports made by civilians on the Twitter platform and detects reports that refer to a potential fire incident. These reports are stored in a structured way. The fire incident report analytic sub-module is responsible for aggregating the detected fire incident reports, and based on the number of these reports and the location these reports refer to, it concludes to a probability that there was a significant amount of people that reported a fire incident at a specific location. The final output is the result along with a geographic area and a reliability score of each location and the coordinates of each location.

3.2. Fire incident detection

3.2.1 Introduction to information extraction

Natural language processing (NLP) is a field of computer science responsible for the study and analysis of raw text. The purpose of this field is to enhance human-computer communication by constructing systems that are capable of understanding raw text and incorporate interaction interfaces based on textual messages. Some of the main topics of NLP are learning syntactic and semantic rules and determining concept, topics, and sentiment from a document, automatic summarization, machine translation, natural language generation, information extraction, etc. [14].

Information extraction corresponds to the section of NLP which is responsible for the analysis of unstructured textual pieces and conversion to a structured form. For example, the conversion of the following unstructured text (raw text):

"Yesterday, New York based Foo Inc. announced their acquisition of Bar Corp."

to the structured form:

MergerBetween('Foo Inc', 'Bar Corp', date ...)

The above-structured form corresponds to a relation of various entities that were embedded in the initial unstructured raw text. The benefit of this conversion is that structured relations can be manipulated by computer algorithms and finally be exploited by computer algorithms. Apparently, for a given unstructured text, many structured forms correspond each one holding different knowledge and representing different relations. As a result, the algorithm designer has the responsibility of selecting the appropriate structured form.

The information extraction procedure consists of the following steps:

1. *Sentence segmentation*: the procedure of distinguishing different sentences.

2. *Tokenization*: the procedure of splitting each sentence to structural components (words and punctuations).

3. *Part of speech tagging*: the procedure of characterizing each token of each sentence to the corresponding part of speech.

4. *Entity recognition*: the procedure of characterizing tokens or set of tokens of each sentence based on previous knowledge. For example, characterize words referring to geographic locations as "city," "country," "mountain," etc.

5. *Relation recognition*: the procedure of detecting specific combination of tokens that corresponds to a specific meaning relation among them. For example, the following segmented tagged sentence **'George' (SUBJECT, NAME)** ↔ **'lives' (VERB, RELEVANT TO LOCATION)** ↔ **'in'** ↔ **'Athens' (OBJECT, LOCATION)** leads to the relation **lives('George', 'Athens')**.

The above procedures make use of text processing algorithms, knowledge representation, and information retrieval algorithms. In order to achieve text segmentation (sentence segmentation or tokenization), each text should be treated as an array of characters. Segmentation is based on the a priori knowledge of special characters that in most cases are used for splitting. For example, sentences usually end with a period mark "." or exclamation mark "!" or question mark "?" and begin with a capital letter. As a result a general rule for segmenting sentences would be to search for pairs: (special character ↔ capital letter) or (special character ↔ end of text).

Tagging procedures are usually based on knowledge databases and information retrieval algorithms. For this task, there is a need of having a lexical and syntactic and semantic database, which we call a corpora (of course different for each language!), which holds characterizations of several words to conceptual entities, and their relations in a structural way. Consequently, segmented texts (tokenized texts) are used as key vectors in order to retrieve from the corpora the corresponding characterization set. The most common approaches for this task are:

- *Sequential classification algorithms*: Hidden Markov Models (HMM) and Conditional Random Fields (CRF).

- *Classification algorithms*: Support Vector Machines (SVM) and Artificial Neural Networks (ANN).

Finally, relation extraction procedures demand from the algorithm designer to predefine either directly by specifying relation rules and use matching algorithms in order to detect word patterns corresponding to specific rule or indirectly by providing to the system several examples of annotated tagged sentences and then use classification algorithms in order to specify the corresponding relations.

3.2.2 Information extraction from Twitter

In this section, a real-case scenario of a system that was realized and evaluated for the purposes of real-time automatic fire detection as demanded by the EU-funded research project "AF3" is presented [15]. The suggested solution comprises a training phase where, via surveys, a variety of tweet samples for various predetermined occasions were collected. These samples were used in order to create a language model (template) that refers to fire incident report.

Training phase: The system presented here is responsible for acquiring reports and comments made by civilians about fire incidents at specific locations. In order to define the algorithms to be used, first it is needed to determine the requirements of these algorithms, the desired performance, and efficiency [16]. Consequently, as a first step, a training comment platform was constructed where users were asked to make some comments about a fire incident that they were witnessed hypothetically (see **Figure 2**). Moreover, they were asked to make some comments that use phrases that refer to fire reports, but the comment *should not* refer to a fire incident but to something else (see **Figure 3**). For example, "John has a burning desire to succeed in his new business" (here "burning" means "very strong").

Figure 2.
Training comment platform: declaration of fire burst.

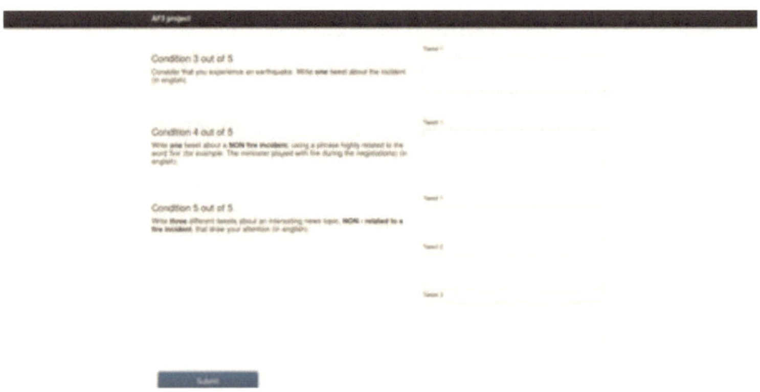

Figure 3.
Training comment platform: tricky "fire" word usage.

3.2.3 Figure training comment platform interface

The results of the training phase were passed through (a) sentence segmentation, (b) tokenization, (c) part of speech tagging, and (d) name entity detection algorithms, so consequently each report was converted to a tagged sentence form:

E.g. <'I'> <'think'> <'there'> <'is', DEFINING VERB> <'fire', FIRE RELATED WORD> <'at'><'Immitos', LOCATION>

As a result, this procedure concluded to a set of tagged sentences that we know that they refer to fire incident report. Next, these reports were aggregated based on their similarity. Finally, the most common aggregated ones were kept in a regular expression form in order to represent the variations. These aggregated rules correspond to the relation rules that will be used by the relation recognition step of the information extraction module. The selected rules are the following:

1. <FIRE RELATED WORD> <EXCLAMATION MARK> * <TIME> + EXCLAMATION MARK> * <VERB LOCATION DEFINITION> + <PREPOSITION> + < HASHTAG> + <LOCATION>

2. <FIRE RELATED NOUN> <EXCLAMATION MARK> * <FIRE RELATED VERB>

3. <FIRE RELATED NOUN> <EXCLAMATION MARK> *<VERB RELATED TO SMOKE> + <PREPOSITION> + <HASHTAG> + <LOCATION>

4. <LOCATION> <EXCLAMATION MARK> * <SENSITIVE AREA> + <FIRE EXPRESSION>

5. <LOCATION> <EXCLAMATION MARK> * <SENSITIVE AREA> + < EXCLAMATION MARK> * <HASHTAG> + <FIRE RELATED NOUN>

6. <LOCATION> <EXCLAMATION MARK> * <FOREST> + <EXCLAMATION> * < FIRE RELATED VERB> <EXCLAMATION MARK> * <HASHTAG> + <FIRE RELATED NOUN>

7. <SENSITIVE AREA> + <EXCLAMATION MARK> * <FIRE EXPRESSION> < EXCLAMATION MARK> * <HASHTAG> + <LOCATION>

where
FIRE-RELATED NOUN: 'fire', 'flames', 'smoke', etc.
VERB LOCATION DEFINITION: verbs that define location ('exists', 'is located', 'is', etc.)

FIRE LOCATION VERB: 'burn', 'fire', etc.
VERB-RELATED TO SMOKE: 'covering', 'smoke', etc.
SENSITIVE AREA: forest, trees, park, etc.

3.3 Fire incident aggregation and potential fire incident prediction

3.3.1 Overview

In the previous section, the procedure of fire incident report acquisition was presented. The result is the gathering of various fire incident reports on different locations with different timestamps. Despite the fact that these reports may seem reliable, due to the severity of the situation, there would be cases, however, that a report may indeed refer to a false fire incident, either because of false fire incident detection from the information extraction component or because of a false report by a civilian [17]. It should be highlighted here that a false report is not made intentionally (like fake news, e.g., as examined in Section 2), but it is an outcome of misunderstanding or a tricky usage of the word fire and its derivatives (i.e., pants on fire). In order to ensure that fire incident notification alerts correspond to a noteworthy event, such reports should be checked of their validity before they are reported to the ingestion server. Consequently, the system consists of an analytic process responsible for the confirmation of the reports based on the number and the location of them. The analytic process implements a reliability model which aggregates the reports and concludes to a fire incident event report along with a reliability score. The reliability score corresponds to the level of how many trustful reports of fire incidents refer to a specific location. The reliability model is presented in more detail in the next section.

3.3.2 Implementation

Initially, the analytic process clusters incident reports based on their geo-coordinates (longitude, latitude). Due to the fact that fire incident reports usually are distributed densely along the fire locations, DBSCAN algorithm [18] was used for report clustering, which is a very efficient dense-based unsupervised classification algorithm for two-dimensional spaces and Euclidean distance as proximity measure and is able to detect accurately various cluster shapes. Then, for each cluster, the reliability model is applied where, finally, a geographical area that it is suspected of being threatened by fire incident is estimated, along with a reliability score.

3.3.3 Reliability model

The reliability model was designed by assuming that very few reports for specific location probably would mean that these reports are probably false alarms, but above a specific threshold, it is almost clear that there is a significant number of people reported a fire incident. In other words if, for example, there emerges one tweet referring to a great fire at the center of Athens, apparently there would be doubts about the validity of this report. Probably, we would say that either this report was a joke or the author of this comment might mean something different that of the literal meaning of a fire incident. On the other hand, if 100 tweets reported a fire incident, probably a real fire incident in the center of Athens is very likely. Apparently, some more tweets would not do the difference. As a result, an exponential model was selected which is parameterized by:

- *Low threshold*: The bottom threshold of the number or reports, where below of it these reports are considered unreliable
- *High threshold*: The upper threshold of the number of reports, where above of it these reports are considered very reliable
- *Low threshold probability (Pl)*: reliability corresponding to the low threshold
- *High threshold probability (Ph)*: reliability corresponding to the high threshold

The reliability score is given by

$$\text{Reliability score} = 1 - b \cdot e^{a \cdot NoR}$$

The term *NoR* stands for the number of results. In case of

$$\text{NoR} = \text{low threshold then we set reliability score} \leftarrow \text{Ph} \quad (1)$$

$$\text{NoR} = \text{high threshold then we set reliability score} \leftarrow \text{Pl} \quad (2)$$

Thus:

$$\text{Eq.}(2) \leftrightarrow 1 - b \cdot e^{a \cdot low \text{ threshold}} = Ph \leftrightarrow \ln((1 - Pl))/b) = a \cdot low\ threshold \quad (3)$$

Similarly:

$$Eq.(3) \leftrightarrow \ln((1 - Ph))/b) = a \cdot high\ threshold \quad (4)$$

$$Eq.(4), Eq.(5) \rightarrow \frac{\ln((1 - Pl))/b)}{\ln((1 - Ph))/b}) = \frac{(\ln(1 - Pl) - \ln(b))}{(\ln(1 - Ph) - \ln(b))} = \frac{Low\ threshold}{High\ threshold} \quad (5)$$

Let:

$$\text{c} = \frac{Low\ threshold}{High\ threshold} \quad (6)$$

Then:

$$Eq.(6), Eq.(7) \rightarrow \ln(1 - Pl) - \ln(b) = c \cdot \ln(1 - Ph) - c \cdot \ln(b) \leftrightarrow b = e^{\frac{(c \cdot \ln(1 - ph) - \ln(1 - Pl))}{c - 1}} \quad (7)$$

Moreover:

$$\text{Eq.}(5), \text{Eq.}(8) \rightarrow a = \frac{1}{High\ threshold} \cdot \ln((1 - Ph))/b) \quad (8)$$

4. Overall proposed scheme for Twitter post-based fire burst detection

Based on the system architectures presented in Sections 2 and 3, we propose a hybrid architecture for detecting fire bursts in real time based on Twitter posts. The proposed architecture can be divided into two parts: a deep learning scheme for distinguishing false from valid Twitter posts and a typical NLP scheme for

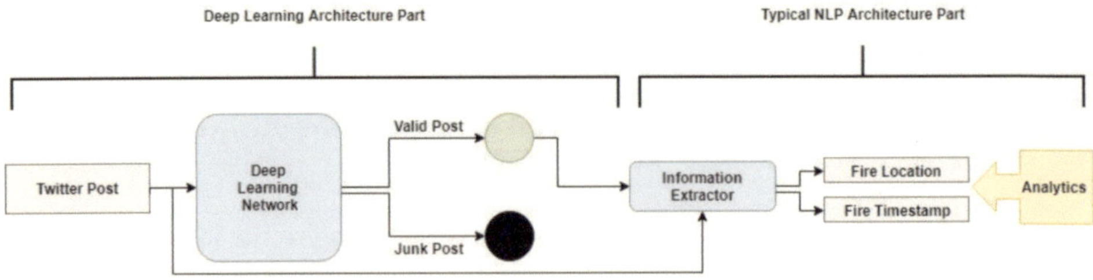

Figure 4.
Proposed overall architecture.

extracting the crucial information with respect to the declared fire burst post. The overall combined scheme is illustrated in **Figure 4**. The deep learning network part represents the scheme presented in Section 2, while the information extractor of the typical NLP part represents the scheme presented in Section 3.

For the fake post detection part, we are to recruit the aforementioned deep learning scheme as it performs twice as good as the related NLP-based methods [19]. Thus, Twitter post processing is expected to work much faster than in the case of implementing a typical NLP-based procedure of the state of the art. In addition, the availability of large posts/news datasets [10–12] facilitates the reliable training of such systems.

Despite the current trend of massively turning to deep neural networks, we designed and constructed a rather typical NLP-based architecture for the information extraction part of our system. This is highly related to the prerequisites that the training procedure of a deep neural network sets, as well as the nature of the problem itself. To begin with, due to lack of a publicly available (i.e., dataset containing a large number of fire burst-related Twitter posts), appropriate dataset for this task, a deep learning approach would be one of only few chances of success. More importantly, the nature of the task itself points to the direction we followed; fire-related posts on a social media platform are reasonably expected to have some common characteristics that make it suitable for a human to model them in order to obtain the desired information. For example, such posts are expected to be short in length, declaring the area of the fire source while containing words and phrases from a fire-related expression set of manageable size. So, our NLP-based subsystem is human and not machine modeled, is proven to be efficient, and is human intuitive and understandable, something that makes it easier to manipulate and expand, if needed.

5. Validation

The system was tested during the AF3 pilot exercise in Skaramagas naval base in two scenarios: (a) fire incident indication based on reports coming from mobile app and (b) fire incident indication based on reports coming from Twitter posts (tweets) containing the hashtag #af3EUprojectFireDetection_TRIAL.

During the first scenario test, a controlled fire was set at an open area inside the naval base. After a while actors, members of the pilot exercise, pretending to be citizens passing by, started posting reports about the fire incident they witnessed. These posts were analyzed by the fire incident detection module and return a notification of a potential of fire incident along with the estimated location and a reliability score. The results were visualized by the public information channel, where fire incident notifications were presented on the map as an area that it was

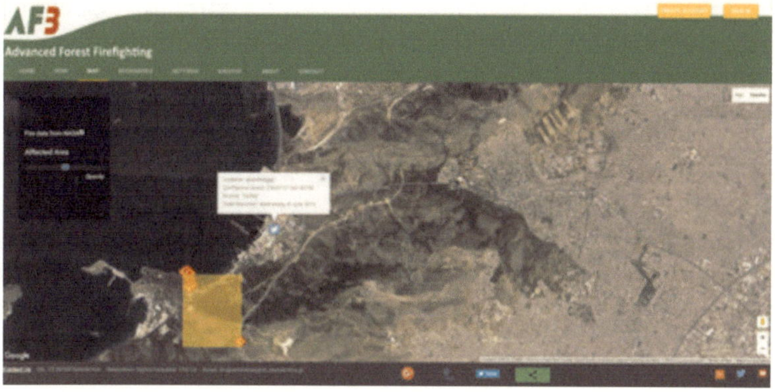

Figure 5.
Validating the scenarios.

Source	Expected results	Validation
Fire incident indication based on reports coming from Twitter posts (tweets)	• Collect the post coming from Twitter and containing the hashtag #af3EUprojectFireDetection_TRIAL • Pass these posts through the information extraction sub-module in order to distinguish the tweets that referred to fire incidents from the ones that did not • Cluster the posts referring to fire incidents, and detect fire incident areas along with the corresponding reliability score • Send result to ingestion server via the REST API	Done successfully

Table 2.
Validation results of the NLP-based scheme.

estimated that the fire was located along with the post comments of the reports, photos attached with the reports, and the reliability score (see **Figure 5**).

During the second scenario test, a controlled fire was set at an open area near the military airport in Aktio [20]. After a while actors, members of the pilot exercise, similarly with the first scenario, committed posts about the fire incident on the Twitter instead of the mobile app. These tweets were collected by the fire incident detection component, analyzed, and distinguished the ones that refer to the fire incident. These reports were gathered by the analytic module and, as described above, clustered, and finally the corresponding notifications were sent to the ingestion server. The results, similar to the first case, were visualized by the public information channel and exploited by the data fusion component in order to enhance its estimation. **Table 2** illustrates the validation results.

6. Conclusions

Fire bursts are a dangerous problem of great importance worldwide. Mega fires often result in significant environmental destructions, major damages on infrastructures, and economic loss. Most importantly, they put at stake the lives, not only of the civilians but also of the forest fire personnel. Thus, technologies that facilitate early fire detection are important for reducing fires and their negative effects.

This chapter aims to provide an alternative view for early fire detection based on twitter posts, instead of expensive sensors and other infrastructures. A hybrid system architecture is introduced which combines a deep learning process for the detection of valid twitter posts regarding fire bursts and a NLP process which extracts the crucial information (place, time, etc.) from the valid tweets. Finally, risk assessment, based on analytics, is performed which derives the geographical places threatened by fire at the current time.

Part of the architecture is already validated under real-world conditions, and the results are promising. The overall system performance is expected to be further improved once the deep learning scheme is entirely utilized.

Acknowledgements

This work was performed within the AF3 Project (Advanced Forest Fire Fighting), with the support of the European Commission by means of the Seventh Framework Programme (FP7), under Grant Agreement No. 607276.

Conflict of interest

There are no "conflict of interest" issues regarding this chapter.

A. Appendices and nomenclature

The mathematical definition of the convolution process between two one-dimensional signals $f(t)$ and $g(t)$ follows in Eq. (9). The mathematics behind LSTM layer architecture follows in Eqs. (10)–(13). Functions σ and *tanh* represent the sigmoid and hyperbolic tangent function, respectively. Parameter W corresponds to weighting matrices:

$$(f * g)(t) = \int_{-\infty}^{+\infty} f(\tau) g(t - \tau) d\tau \tag{9}$$

$$z_t = \sigma(W_z \cdot [O_{t-1}, O_t]) \tag{10}$$

$$r_t = \sigma(W_r \cdot [O_{t-1}, O_t]) \tag{11}$$

$$h'_t = \tanh(W \cdot [r_t * O_{t-1}, O_t]) \tag{12}$$

$$O_t = (1 - z_t) * O_{t-1} + z_t * h'_t \tag{13}$$

Author details

Konstantinos-George Thanos*, Andrianna Polydouri, Antonios Danelakis, Dimitris Kyriazanos and Stelios C.A. Thomopoulos
Integrated Systems Laboratory (ISL), Institute of Informatics and Telecommunications, National Center for Scientific Research "Demokritos", Athens, Greece

*Address all correspondence to: giorgos.thanos@iit.demokritos.gr

References

[1] Titcomb, J, Carson, J. Fake News: What Exactly Is It—And How Can You Spot It? [Internet]. 2018. Available from: https://www.telegraph.co.uk/technology/0/fake-news-exactly-has-really-had-influence/ [Accessed: 16 January 2018]

[2] Singhania S, Fernandez N, Rao S. 3HAN: A deep neural network for fake news detection. In: Proceedings of the International Conference on Neural Information Processing (ICONIP 2017); 14-18 November 2017; Guangzhou, China. New York: Springer; 2018. pp. 572-581

[3] Wang WY. Liar, liar pants on fire. A new benchmark dataset for fake news detection. Computation and Language. 2017, arXiv preprint: 1-5. Available from: arXiv:1705.00648

[4] Shu K, Mahudeswaran D, Liu H. FakeNewsTracker: A tool for fake news collection, detection, and visualization. Computational and Mathematical Organization Theory. 2018:1-12. https://link.springer.com/article/10.1007/s10588-018-09280-3

[5] Hochreiter S, Schmidhuber J. Long short-term memory. Neural Computation. 1997;**9**(8):1735-1780

[6] Popat K, Mukherjee S, Yates A, Weikum G. DeClarE: Debunking fake news and false claims using evidence-aware deep learning. Computation and Language. 2018, arXiv preprint: 1-11. Available from: arXiv:1809.06416

[7] Oluwaseun A, Deepayan B, Shahrzad Z. Fake news identification on Twitter with hybrid CNN and RNN models. In: Proceedings of the 9th International Conference on Social Media and Society (SMSociety '18); 18-20 July 2018; Copenhagen, Denmark. New York: ACM; 2018. pp. 226-230

[8] Zhang J, Cui L, Fu Y, Gouza FB. Fake news detection with deep diffusive network model. Social and Information Networks. 2018, arXiv preprint: 1-10. Available from: arXiv:1805.08751

[9] Goodfellow I, Bengio Y, Courville A. Deep Learning. Cambridge, Massachusetts, USA: MIT Press; 2016. pp. 326-716. Available from: http://www.deeplearningbook.org/ [Accessed: 23 January 2019]

[10] Gillin J. Politifact's Guide to Fake News Websites and What They Peddle [Internet]. 2017. Available from: https://www.politifact.com/punditfact/article/2017/apr/20/politifacts-guide-fake-news-websites-and-what-they/ [Accessed: 18 January 2019]

[11] University of Warwick. PHEME rumor dataset: Support, certainty and evidentially [Internet]. 2016. Available from: https://www.pheme.eu/2016/06/13/pheme-rumour-dataset-support-certainty-and-evidentiality/ [Accessed: 18 January 2019]

[12] Zubiaga A, Liakata M, Procter P, Wong Sak Hoi G, Tolmie P. Analysing how people orient to and spread rumours in social media by looking at conversational threads. PLoS One. 2016; **11**:1-29. DOI: https://doi.org/10.1371/journal.pone.0150989

[13] Hsu ST, Moon C, Jones P, Samatova N. A hybrid CNN-RNN alignment model for phrase-aware sentence classification. In: Proceedings of the 15th Conference of the European Chapter of the Association for Computational Linguistics (EACL 2017); 3-7 April 2017; Valencia, Spain, Short Papers. Vol. 2. 2017. pp. 443-449

[14] Bird S, Klein E, Loper Ed. Natural Language Processing with Python. 1st ed. Sebastopol, California, USA: O' Reilly Media; 2009. Available from: http://www.nltk.org/book [Accessed: 23 January 2019]

[15] AF3 EU-Project. Advanced Forest FireFighting [Internet]. Available from: http://af3project.eu/ [Accessed: 23 January 2019]

[16] Imran M, Castillo C, Diaz F, Vieweg S. Processing social media messages in mass emergency: A survey. Social and Information Networks, 2015, arXiv preprint: 1-37. Available from: arXiv: 1407.7071

[17] Mendoza M, Poblete B, Castillo C. Twitter unser crisis: Can we trust what we RT? In: Proceedings of the First Workshop on Social Media Analytics (SOMA 2010); 25 July 2010; Washington DC, USA. Pennsylvania Plaza, New York City, USA: ACM; 2010. pp. 71-79

[18] Ester M, Kriegel HP, Sander J, Xu X. A density-based algorithm for discovering clusters in large spatial databases with noise. In: Proceedings of the Second International Conference on Knowledge Discovery and Data Mining (KDD 1996); 2-4 Aug 1996; Oregon, USA. Vol. 969(34). 1996. pp. 226-231

[19] Oshikawa R, Qian J, Yang Wang W. A survey on natural language processing for fake news detection. Computation and Language, 2018, arXiv preprint: 1-11. Available from: arXiv:1811.00770

[20] Thomopoulos CAT, Kyriazanos DM, Astyakopoulos A, Lampropoulos V, Dimitros K, Margonis C, et al. OCULUS fire: A control and command system for fire management with crowd sourcing and social media interconnectivity. In: Proceedings of SPIE Defence, Security and Sensing (SPIE DSS 2016); 17-21 April 2016; Baltimore, Meryland, USA. Vol. 9842. 2016. p. 98420U

Chapter 7

Blind Wavelet-Based Image Watermarking

Abeer D. Algarni and Hanaa A. Abdallah

Abstract

In this chapter, the watermarking technique is blind; blind watermarking does not need any of the original images or any information about it to recover watermark. In this technique the watermark is inserted into the high frequencies. Three-level wavelet transform is applied to the image, and the size of the watermark is equal to the size of the detailed sub-band. Significant coefficients are used to embed the watermark. The proposed technique depends on quantization. The proposed watermarking technique generates images with less degradation.

Keywords: watermarking, discrete wavelet transform, quantization, blind, coefficients, peak signal-to-noise ratio, normalization, correlation

1. Introduction

Watermarking methods operating in the wavelet domain have become attractive because they have inherent robustness against compression if the low-frequency band is selected for watermark embedding, and, additionally, the wavelet transform provides a multiresolution representation of images, which can be exploited to build more efficient watermark detection schemes. The history of watermarking is presented here. Zhu et al. [1] proposed adding a mark, a Gaussian sequence of pseudorandom real numbers, into all the high-pass bands in the wavelet domain. An algorithm developed by Xia et al. [2] utilizes large DWT coefficients of the high- and mid-frequency bands to embed a random Gaussian distributed watermark sequence. Dugad et al. [3] provided a method to embed a Gaussian sequence of pseudorandom real numbers into selected coefficients in all detailed sub-bands with magnitude above a given threshold in a three-level decomposition with Daubechies-8 filters. In general, the watermark embedded in low-pass bands of the wavelet domain is robust to a group of attacks such as low-pass filtering, Gaussian noise, and lossy compression but affects the fidelity of the watermarked image and that in high-pass bands is resistant to another set of attacks such as histogram equalization, intensity adjustment, and gamma correction [4].

2. Blind and non-blind methods

As described before, watermarking methods can be classified according to whether the original data is used in extraction/detection procedure or not. In 1997, Cox et al. [5] proposed a watermarking method where they embed the watermark into the lower frequency coefficients in the DCT domain. Their method needs the original image and the embedding strength coefficient to detect the presence of the watermark. However, the original source might not be available in several applications. Barni et al. [6] presented a method to overcome the non-blind watermarking problem. They correlate the watermark sequence directly with all coefficients of the received image and then compare the correlation coefficient with some detection threshold. Only, the watermark sequence and the scaling factor are needed in the watermark detection. This approach is widely utilized in the watermarking community. However, it turns out that blind methods are less secure than non-blind methods.

3. Watermarking in transform domains

Watermarking methods can be classified according to whether they use embedding based on additive algorithms or quantization algorithms.

3.1 Additive algorithms

Additive embedding strategies are characterized by the linear modification of the host image and correlative processing in the detection stage. A considerable number of image watermarking methods share this architecture. In most algorithms, the signature data is a sequence of numbers wi of length N that is embedded in a suitable selected subset of the host signal coefficients. The basic and commonly used embedding formulas are defined by the following equations (Eqs. (1) and (2)):

$$V_i' = V_i(1 + k.w_i) \quad (1)$$

$$V_i' = V_i + k.w_i \quad (2)$$

where *k* is a weighting factor that influences the robustness as well as the visibility and V' is the resulting modified host data coefficients carrying the watermark information. The majority of watermarking systems presented in the literature falls into this class, differing chiefly in the signal design, the embedding, and the retrieval of the watermark content. The extraction process is accomplished by applying the inverse embedding formulas.

The algorithm developed by Dugad et al. [3] makes use of a sequence of pseudorandom Gaussian real numbers, matching the size of the detailed sub-bands of the wavelet domain. The authors performed three-level decomposition with Daubechies-8 filters and selected all coefficients in all detailed sub-bands, whose magnitude is above a given threshold. The equation used for watermark embedding is described in Eq. (3).

For a blind retrieval of the watermark, a statistical detector was proposed based on the following formula:

$$\delta = \frac{\sum_N V_i^* .w_i}{N} \quad (3)$$

where δ is estimated by correlating the watermark sequence w directly with all N coefficients of the received image V^*. A large number of random sequences are tested, but only the sequence that was originally embedded yields the highest correlation coefficient. Therefore, we can conclude that the image has been watermarked with w. A detection threshold τ can be established to make the detection decision if δ τ. The detection threshold can be derived either experimentally or analytically.

The threshold τ is estimated using Eq. (4):

$$\tau = \frac{\alpha}{2.N} \sum_{i=1}^{N} \left| V_i^* \right| \tag{4}$$

where only the coefficients above the detection threshold are considered.

3.2 Algorithms based on quantization

The quantization schemes perform nonlinear modifications during embedding and detecting the embedded message by quantizing the received samples to map them to the nearest reconstruction point. Quantization is the process of mapping a large possibly infinite set of values to a much smaller set. A quantizer consists of an encoder mapping and a decoder mapping. The range of source values is divided into a number of intervals. The encoder represents each interval with a code word assigned to that interval. The decoder is able to reconstruct a value for every code word produced by the encoder. Scalar quantizes take scalar values as input and output code words, while vector quantizers work with vectors of input sequences or blocks of the source input.

Quantization-based watermarking is a new technique, as a logo is embedded and detected in a blind way. Authors in [6] introduced a scalar quantization watermarking technique, where the watermark is embedded in the middle- and low-frequency bands. The robustness of the algorithm is tested by applying the algorithm to JPEG compression. Only this attack is tested.

Authors in [7, 8] present another quantization-based watermarking algorithm which improves on the Tsai algorithm by incorporating variable quantization and resistance against a wide range of attacks like blurring, noising, sharpening, scaling, cropping, and compression.

The main issue with these quantization-based algorithms is that it only tackles a subset of attacks. For example, Tsai's algorithm is only robust against JPEG compression; however Chen's algorithm does not tackle geometric attacks like rotation. Hence we propose a new algorithm which is robust against cropping, JPEG compression, resizing, rotation, and salt and pepper.

4. Wavelet-based methods

The wavelet transform finds a great popularity in the field of watermarking as it is able to decompose the available images into sub-bands, in which watermarks can be embedded [3, 9]. Taking the cue from the spread spectrum method, we embed the data in transform coefficients chosen in a random order. For extraction of the hidden data, the random sequence must be made available to the extractor. Cox et al. [5] were the first to apply the spread spectrum method to data hiding. Transforms such as the DCT and DWT have been used. The use of the DWT has advantages of speed and robustness against wavelet-based compression. Previously, Dugad's algorithm introduced an additive watermarking technique in the wavelet domain [3]. The proposed technique in this paper uses three-level wavelet transform using Daubechies filter; the watermark is embedded in the high-frequency domain [9], and it is blind algorithm, and the watermark is detected without using the original image. Also this technique uses only the high value coefficients to insert the watermark. Large wavelet coefficients are referred to edges within an image. So, any degradation in this region won't be noticed by the human viewer. Also it is difficult to remove the watermark without distorting the marked image according to the perceptually significant large magnitude wavelet coefficients. Since watermark verification typically consists of a correlation estimation step, which is extremely sensitive to the relative order in which the watermark coefficients are placed within the image, such changes in the location of the watermarked coefficients were unacceptable. Dugad et al. have proposed a spread spectrum method for digital image watermarking in the wavelet domain, which does not require the original image for watermark detection [3]. This method is based on adding the watermark in selected coefficients with significant image energy in the transform domain in order to ensure non-erasability of the watermark. This method has an advantage over the previous methods, which did not use the original in the detection process and could not selectively add the watermark to the significant coefficients, since the locations of such selected coefficients can change due to image manipulations.

The method proposed by Dugad et al. [3] has overcome the problem of "order sensitivity." It has some advantages such as an improved resistance to attacks on the watermark, an implicit visual masking utilizing the time-frequency localization property of the wavelet transform, and a robust definition for the threshold, which validates the watermark.

The disadvantage of this method is using additive technique in watermarking. In this additive method, the detectors must correlate watermarked image coefficients with the known watermark to know if the image is marked or not. To solve this problem, it is important to correlate a large number of coefficients as possible, but it in turn requires the watermark to be embedded into many image coefficients at the embedding stage. This has the effect of increasing the amount of degradation in the marked image. Another drawback is that the detector can only tell if the watermark is present or absent. It cannot recover the actual watermark. Here, we present a new method to avoid these drawbacks. It is possible to use the advantages of the watermarking scheme by Dugad et al. [3] while avoiding the disadvantages. This can be achieved using the idea of a watermark with the same size as the original image in conjunction with adapted versions of scalar quantization insertion/detection method. The resultant watermarking system will be blind and based on quantization.

A watermark size has to be equal in size to the detailed sub-band in wavelet transform domain, and only significant coefficients will be used to embed watermark. Finally, this new method outperforms the previous method using quantization and a new watermark embedding process, not the additive one. After applying a comparable robustness performance, the watermarked images using our new method give less degradation than Dugad's scheme.

However, only a few of these watermark values are added to the host image. The watermark values are found in fixed locations; thus, the ordering of significant coefficients in the correlation process is not an issue for watermark detection. This gives the technique a value as the correlation process is sensitive to the ordering of significant coefficients, and if there is any change applied to the ordering, it will cause a poor detector response.

In Zolghadrasli's method that is based on the DWT [10], Gaussian noise is used as the watermark. Here the watermark is added to the significant coefficients of each selected sub-band depending on the human visual system (HVS) characteristics. Any small modifications are performed to improve HVS model. This technique is non-blind as the host image is needed in the watermark extraction.

4.1 Dugad's method

Dugad et al. [3] presented an additive watermarking method operating in the wavelet domain. This method allowed the detection of the watermark without access to the original uncorrupted image.

4.1.1 Embedding algorithm

The embedding algorithm can be summarized in the following steps:

1. From all wavelet coefficients (except the low-pass coefficients in LL band and high-pass coefficients in HH band), the coefficients of magnitude higher than t1 are chosen. This proves that only significant coefficients are used. The wavelet coefficients of magnitude higher than t1 depend upon the smoothness or more details in the image.

2. Then the zero mean and unit variance watermark are generated with a known seed value; the watermark should be equal in size to the input image.

3. The watermark is embedded in each location which has wavelet coefficient with magnitude higher than t1; the watermarked wavelet coefficient is given by Eq. (5):

$$\hat{w}_{ij} = w_{ij} + k\left|w_{ij}\right|x_{ij} \tag{5}$$

where w_{ij} is the wavelet coefficient, k is a scaling parameter, x_{ij} is a watermark value, and is the watermarked wavelet coefficient.

4.1.2 Detection algorithm

1. The watermark is regenerated using the known seed value.

2. All wavelet coefficients (barring the LL and HH components) of magnitude greater than t2 from a possibly corrupted watermarked image are selected. Note that by setting t2 > t1, we find that the robustness is increased, and some wavelet coefficients with magnitudes below t1 may become higher than t1 due to image manipulations.

3. Wavelet coefficients with magnitude higher than t2 are used in the detection process; these detected values are correlated with the watermark values at the same locations. After this correlation process, a yes or no answer will be given as to the presence of the watermark.

5. Non-blind watermarking

Another watermarking method operating upon significant coefficients within the wavelet domain was presented by Miyazaki et al. [9]. This method takes a three-level wavelet transform of the image to be watermarked and inserts the watermark into the detail coefficients at the coarsest scales (LH3, HL3); the low-pass component LL3 and diagonal details HH3 are excluded.

5.1 Miyazaki's method

Two watermarking algorithms were presented by Miyazaki et al. [9]. Both algorithms were implemented in the wavelet domain, but each targeted a different set of coefficients for insertion. The first of these insertion methods is applied on insignificant coefficients, whereas the second type of insertion is applied on significant coefficients. So, both insertion techniques would be applied to a single image at the same time. However, experimental results proved that the insertion method by applying the significant coefficients was more robust than the insertion method using insignificant coefficients. So, the insertion method utilizing the significant coefficients will be considered.

In this technique three-level wavelet transform is applied to the image, and the watermark is inserted into the detail coefficients at the wavelet level three. The detailed coefficients which are found at level three are the horizontal details, High Low 3 (HL3); the vertical details, Low High 3 (LH3); and the diagonal details, High High 3 (HH3). The low-pass component, Low Low 3 (LL3), is left unchanged. This is a quantization-based watermarking method which aims to modify wavelet coefficients of high magnitude, thus embedding the watermark into edge and textured regions of an image. The process for watermark insertion is as follows:

5.1.1 Embedding algorithm

1. Two thresholds, t1 and t2, are selected, and any one of the sub-bands LH3 and HL3 is chosen. Next, significant coefficients Cok (k = 1, 2... N) satisfying t1 < Cok < t2 are found.

2. A binary watermark is created, Wat (k); k = 1, 2... N.

3. For k = 1, 2,... N, the embedding of the watermark is applied by modifying Ck as follows:

   If Wat(k) = 1 and Cok > 0, then Cok = t2,
   If Wat(k) = 0 and Cok > 0, then Cok = t1,
   If Wat(k) = 1 and Cok < 0, then Cok = −t2,
   If Wat(k) = 0 and Cok < 0, then Cok = −t1,

4. The embedded position, sub-band label, and the two thresholds t1 and t2 should be saved.

5.1.2 Detection algorithm

The following process details the steps involved for watermark detection:

1. Using the sub-band label and the embedded position, the recovered wavelet coefficients Cok. k = 1, 2... N are obtained.

2. Check each Ck individually:

   If Cok < (t1 + t2) /2, then the recovered watermark bit is 0.
   If Cok ≥ (t1 + t2) /2, then the recovered watermark bit is 1.

This thesis introduces a new quantization-based, blind watermarking algorithm operating in the wavelet domain. This algorithm has several advantages as compared to previously published algorithms. For example, the proposed algorithm is better than the algorithm of Dugad in its ability to survive the same malicious attacks while producing marked images of greater visual quality. The proposed watermarking scheme is a blind scheme not requiring a file containing the positions of the marked coefficients as in the method of Miyazaki.

6. The proposed watermarking scheme

The proposed watermarking scheme is a blind quantization-based scheme. A block diagram detailing its steps is shown in **Figure 1**.

6.1 Watermark embedding

1. The cover image is decomposed into sub-bands using three levels of Daubechies wavelet transform using filters of length 4.

2. Then the coefficients in the third level (except the LL3 and HH3 sub-bands) which have magnitude higher than t1 and lower than t2 are chosen to hide in. Let be the wavelet coefficient with maximum absolute in both HL3 and LH3 sub-bands. A threshold $t = \alpha$ is selected, as mentioned in Eq. (6):

$$0.01 < \alpha < 0.1 \text{ and } t2 > t1 > t. \tag{6}$$

3. Then the binary watermark is created using a secret key, which is a seed of a random generator; the watermark size should be of the same size as the two sub-bands which are selected for embedding.

4. Then apply quantization to each of the selected wavelet coefficients.

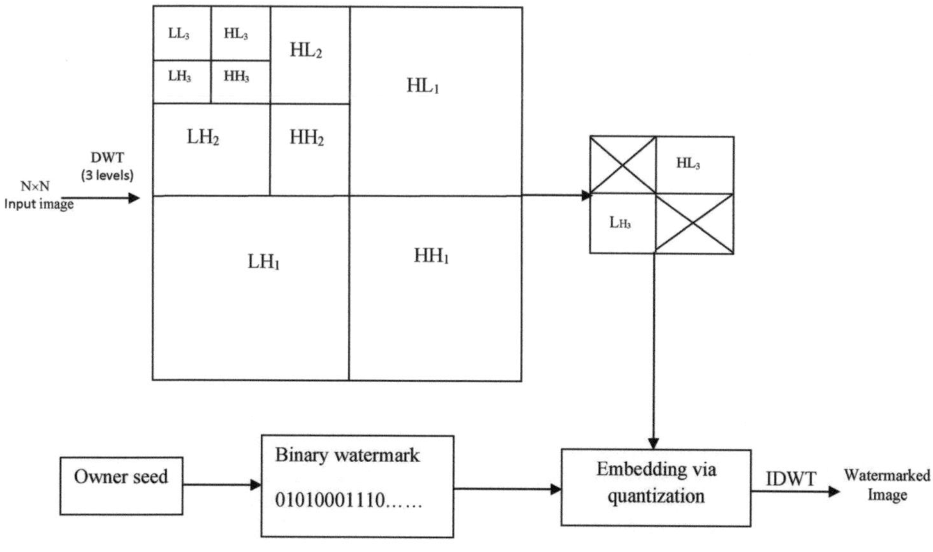

Figure 1.
The proposed image watermarking scheme.

The quantization process is done as shown in Eq. (7):

$$\begin{aligned} &\text{If} = 1 \text{ and} >0, \text{then} = t2 - X1, \\ &\text{If} = 0 \text{ and} >0, \text{then} = t1 + X1, \\ &\text{If} = 1 \text{ and} < 0, \text{then} = -t2 + X1, \\ &\text{If} = 0 \text{ and} < 0, \text{then} = -t1 - X1, \end{aligned} \tag{7}$$

where is the watermark bit corresponding to, and is the watermarked wavelet coefficient. The parameter x1 narrows the range between the two quantization levels t1 and t2 in order to perform a robust oblivious detection. **Figure 2** shows the watermark embedding in a positive wavelet coefficient.

5. After all the selected coefficients are quantized, the inverse discrete wavelet transform (IDWT) is applied, and the watermarked image is obtained.

6.2 Watermark detection

1. The possibly corrupted watermarked image is transformed into the wavelet domain using the same wavelet transform as in the embedding process.

2. The extraction is performed on the coefficients in the third wavelet level (excluding the LL3 and HH3 sub-bands).

3. All the wavelet coefficients of magnitude higher than or equal to t1 + X2 and less than or equal to t2 − X2 are chosen, which are named $w_{ij}^{\prime s}$. Note that the value of X2 should be lower than the value of X1. This maintains that all the marked coefficients are recovered and dequantized after being attacked. The determination of parameters X1 and X2 to the watermarking technique gives a

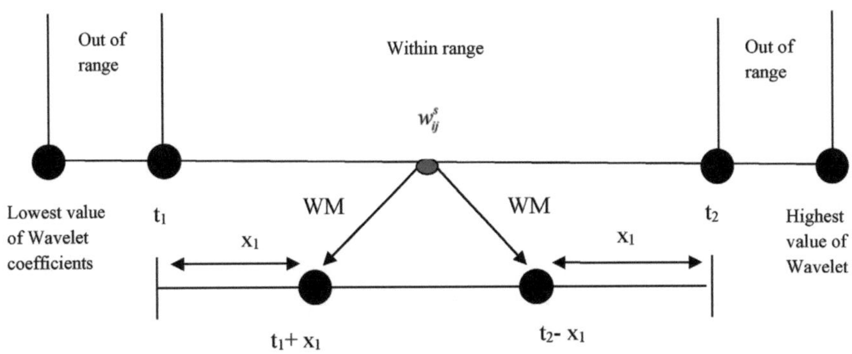

Figure 2.
Watermark embedding for wavelet coefficients in the proposed scheme.

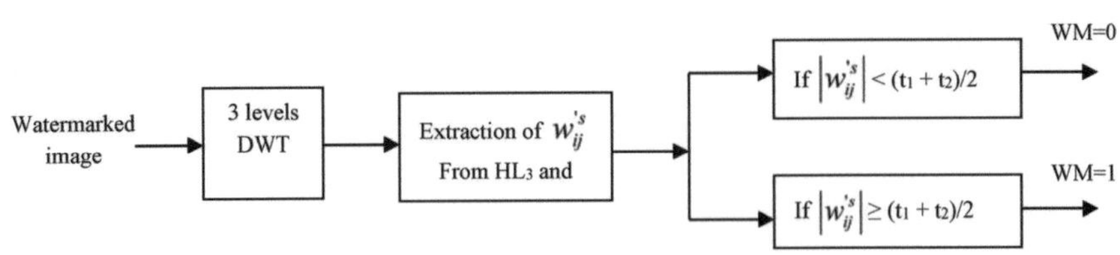

Figure 3.
Watermark detection in the proposed scheme.

degree of tolerance to the system against attacks, i.e., the extraction of watermark bits from the selected wavelet coefficients is done using Eq. (8).

$$\begin{aligned} &\text{If} < (t1 + t2)/2, \text{ the recovered watermark bit is } 0. \\ &\text{If} \geq (t1 + t2)/2, \text{ the recovered watermark bit is } 1 \end{aligned} \tag{8}$$

The watermark detection process can be shown in **Figure 3**.

Then the correlation process is applied between the recovered watermark and the original watermark, obtained via the secret key, just only in the locations of the selected coefficients.

7. The histogram of equal-area division quantization method in watermarking

The quantization levels are calculated using a method dependent on the image content, and then round off the value of pixels to the nearest quantization level. Using this method, the number of values transmitted over the channel is minimized. HEAD is a quantization method in which the transmitted values are reduced by mapping the values of image pixels to a finite number of quantization levels.

Process of HEAD quantization [11]:

1. First of all, get the histogram of the output, and then the area under the histogram is divided into a number of vertical slices with equal areas. A width of each slice is inversely proportional to its height. Quantization levels are determined by the number of these slices. Both are equal.

2. The midpoint value which is found on the width of each slice is considered as a quantization level.

3. This is called a nonuniform quantization where the density of the quantization levels increases with increasing the probability of the occurrence of the pixel value.

4. We mapped all the pixel values that lie within the width of a slice to the quantization level that is represented by the midpoint of this slice.

7.1 Proposed DWT-HEAD watermarking method

7.1.1 Watermark embedding

The steps of watermark embedding can be summarized as follows:

1. The host image is transformed into the wavelet domain; three-level Daubechies wavelet with filters of length 4 is used. The coefficients of HL3 coefficients are watermarked using HEAD quantization using two quantization levels t1 and t2.

2. Each of the selected wavelet coefficients is quantized. After all the selected coefficients are quantized, the inverse discrete wavelet transform is applied, and the watermarked image is obtained.

7.1.2 Watermark detection

1. The possibly corrupted watermarked image is transformed into the wavelet domain using the same wavelet transform as in the embedding process.

2. The extraction is performed on the coefficients in the first level wavelet transform (HL1).

3. All the wavelet coefficients of magnitude greater than or equal to t1 and less than or equal to t2 are selected. The watermark bits are extracted from each of the selected DWT coefficients with Eq. (9):

$$\begin{aligned}&\text{If} < (t1 + t2)/2, \text{ then the recovered watermark bit is } 0.\\&\text{If} \geq (t1 + t2)/2, \text{ then the recovered watermark bit is } 1.\end{aligned} \tag{9}$$

8. Simulation results

This section presents the results to compare between the schemes of LSB method, Dugad's method, Miyazaki's method, and the proposed method. Several images are watermarked using the four watermarking methods and subjected to attacks. In order to measure the degradation suffered by host images after watermark insertion, the PSNR is used.

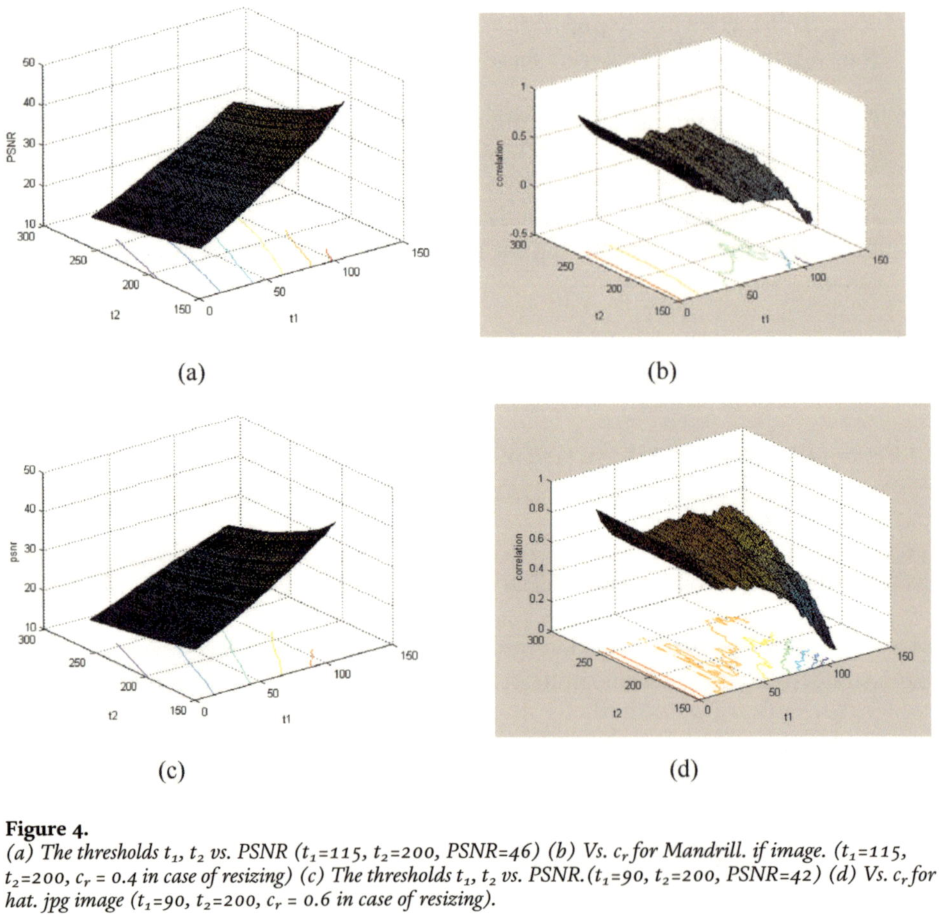

Figure 4.
(a) The thresholds t_1, t_2 vs. PSNR (t_1=115, t_2=200, PSNR=46) (b) Vs. c_r for Mandrill. if image. (t_1=115, t_2=200, c_r = 0.4 in case of resizing) (c) The thresholds t_1, t_2 vs. PSNR. (t_1=90, t_2=200, PSNR=42) (d) Vs. c_r for hat. jpg image (t_1=90, t_2=200, c_r = 0.6 in case of resizing).

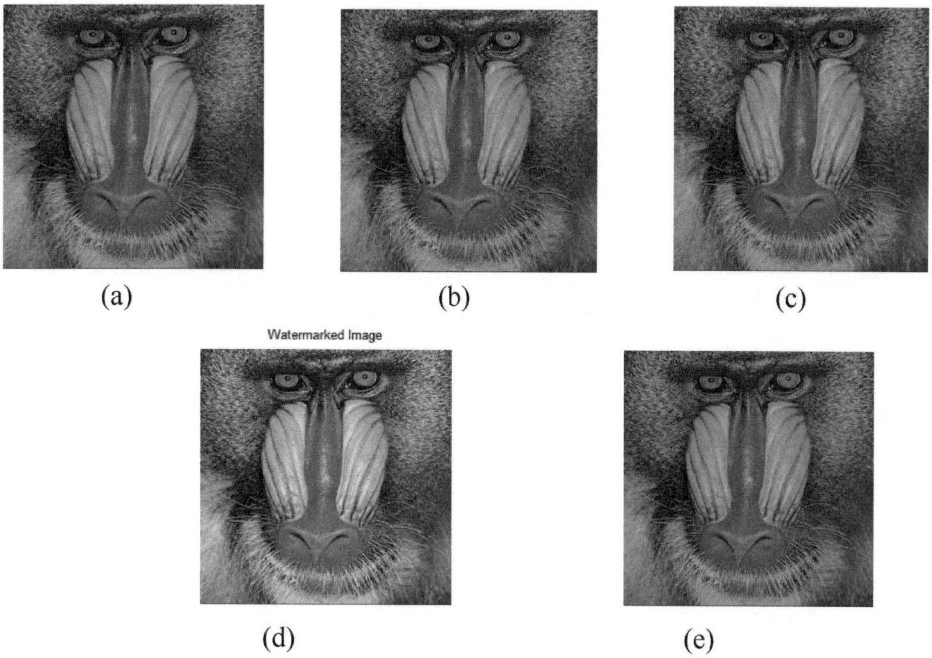

Figure 5.
(a) Original image. (b) Mandrill image marked using watermarking scheme of Dugad in the absence of attacks. (c) Hat image marked using watermarking scheme of Miyazaki in the absence of attacks. (d) Mandrill image marked using LSB. (e) Mandrill image marked using the proposed watermarking method in the absence of attacks.

For all the tests in this chapter, MATLAB is used. All tests are performed upon the 8-bit grayscale 256 × 256 Mandrill, Hat, and Lena images. To simulate the watermarking schemes on the Mandrill image, we set t1 = 115, t2 = 200, and k = 0.1. The suitable thresholds are obtained from the curves in **Figure 4b**. The watermarked images are then attacked with JPEG compression with different compression ratios to make the quality of the images at levels 5 (Q5), 10 (Q10), and 15 (Q15) at the JPEG standard. Other attacks such as the additive white Gaussian noise (AWGN) and cropping attacks are also considered. The same schemes are also applied to the Hat image with similar attacks. The thresholds used for this case are t1 = 90 and t2 = 200. We find from the figures that the suitable thresholds are coming from the curves in **Figure 4d**. To investigate the watermarking methods, we calculate the threshold (t) by using fmax= 528.4 and k = 0.1 so the threshold t = 0.1* fmax=52.84, we will use t1 = 90, t2 = 200 that give the tradeoff between PSNR and correlation as shown in **Figure 4**. The attacks were used to test the new algorithm, we choose the thresholds according to that gives the trade off between the high PSNR and the high Correlation ,in the case of mandrill we find that t1 = 115, t2 = 200, X1 = 20 and X2 = 10. **Figure 5** shows this Watermarked image and the effect of attacking this watermarked image with various attacks. The watermarked images are then attacked with JPEG at levels Q5, Q10, and Q15, AWGN, and cropping.

It can be seen that the watermarking algorithm of Dugad is surviving all the attacks. The high compression ratio using JPEG with quality 5 is one of the attacks applied to the watermarked image and resizing from 256 to 128 is the other attack, it is found that the watermark was not always detected. Results are shown in **Tables 1–3**.

Similar experiments and attacks are carried out for the algorithm in Miyazaki method with t1 = 115 and t2 = 200; we find that the results are better than that of Dugad method because it is a semi-blind method. Results are shown in **Tables 1** and **3**.

Scheme	PSNR	NC
LSB blind	49.9	1
Dugad's blind	42.48	0.57
Miyazaki's non-blind	44.65	1
Proposed scheme blind	46.60	1
Using t_1 = 115, t_2 = 200, and k = 0.1.		

Table 1.
Comparing the proposed method with the other three methods of Dugad, Miyazaki, and LSB (Mandrill image).

Types of attacks	NC	WM length in	WM length out
No attacks	1	102	102
JPEG Q5	0.14	102	53
JPEG Q10	0.48	102	77
JPEG Q15	0.85	102	79
Gaussian (0.006)	0.54	102	54
Salt and pepper (0.15)	0.79	102	79
Cropping	0.48	102	38
Half sizing	0.39	102	48

Table 2.
Comparing NC value for the proposed method with the methods of Dugad, Miyazaki, and LSB.

NC				
Type of attacks	**Blind LSB**	**Blind scheme of Dugad**	**Non-Blind scheme of Miyazaki**	**Blind proposed method**
JPEG Q5	0.0111	0.57	0.75	0.14
JPEG Q10	0.01	0.24	1	0.48
JPEG Q50	0.0193	0.22	1	0.85
Gaussian 0.006	0.0052	0.52	1	0.57
Gaussian 0.01	0.011	0.19	0.93	0.4
Gaussian 0.1	0.0134	0.01	0.49	0.06
Salt and pepper 0.015	0.0030	0.53	0.87	0.45
Salt and pepper 0.15	0.0017	0.037	0.55	0.36
Salt and pepper 0.5	0.0027	0.001	−0.01	0.29
Cropping	−0.0054	0.58	0.95	0.39
Half sizing	0.4245	0.17	0.77	0.49
Subsample 0.7	0.5608	0.25	0.49	0.223
Subsample 0.4	0.3098	0.16	0.71	0.2

Table 3.
Results for the proposed scheme (Mandrill image) (with t_1 = 115, t_2 = 200, X_1 = 20, and X_2 = 10).

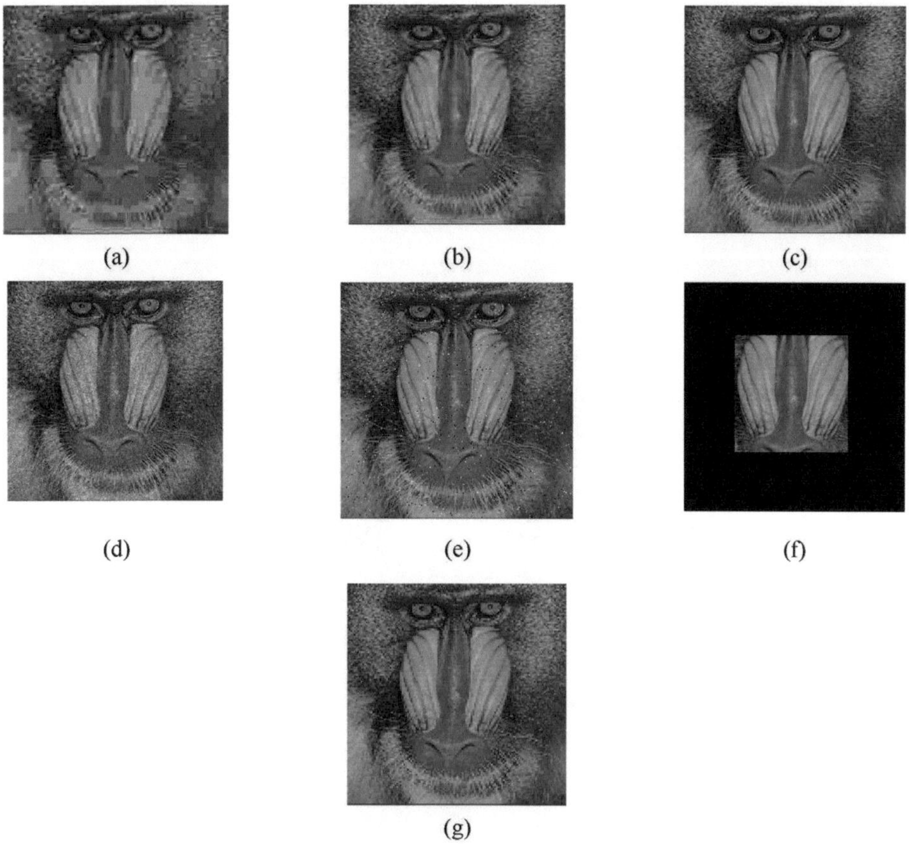

Figure 6.
Attacked image with (a) JPEG quality 5, (b) JPEG quality 10, (c) JPEG quality 15, (d) Gaussian noise (variance = 0.0058), (e) impulse noise (normalized density of 0.015), (f) cropping, and (g) half sizing (followed by resizing back to the original size).

The same attacks were used to test the new algorithm; The thresholds are chosen carefully to achieve tradeoff between the high PSNR and the high Correlation. In the case of Mandrill, we found that t1 = 115, t2 = 200, X1 = 20, and X2 = 10. **Figure 6** shows this watermarked image and the effect of attacking this watermarked image with various attacks. **Table 3** presents the quantitative results for these various attacks.

However, the "cropping" attack poses a problem in that only 38 out of a possible 102 watermark bits were used by the detector, thus decreasing the reliability of the scheme. The scheme is not robust to JPEG quality 5 attack (just like the Dugad method). Thus, while surviving the same attacks as the Dugad scheme, the new scheme does not degrade the watermarked image to the same extent. From **Table 1**, PSNR value is 42.48 dB. The PSNR recorded for the Miyazaki scheme is equal to 44.65dB, the recorded PSNR for LSB is (49.9dB) and PSNR recorded for the new scheme is (46.60dB).

Similar experiments and attacks are carried out for the algorithm in Miyazaki method, Dugad method, LSB method, and the proposed method on Hat image and Lena image, and the results are shown in **Figures 7–10**.

Table 4 presents the PSNR and NC for the proposed method and the other two methods using Hat image. It is seen that our method does not degrade the watermarked image to the same extent as the other two methods. **Table 5** represents the NC for the attacked watermarked images in our proposed method and the other existing methods.

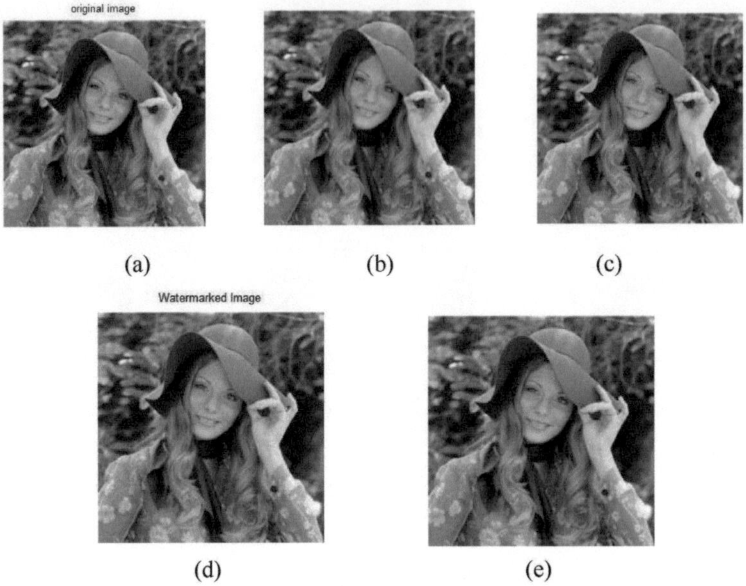

Figure 7.
(a) Original image. (b) Hat image marked using watermarking scheme of Dugad in the absence of attacks. (c) Hat image marked using watermarking scheme of Miyazaki in the absence of attacks. (d) Hat image marked using LSB scheme. (e) Hat image marked using the proposed watermarking method in the absence of attacks.

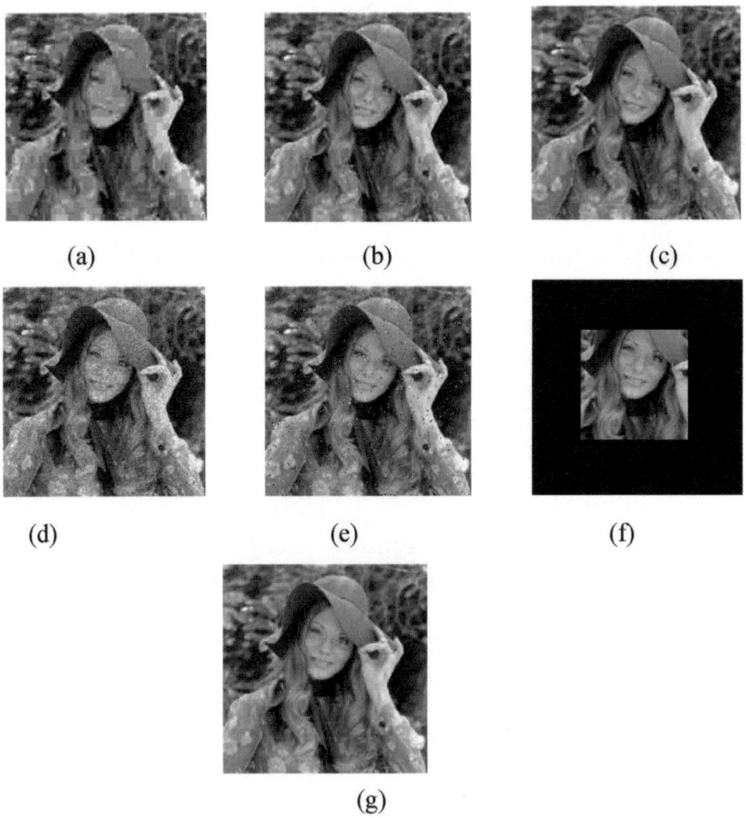

Figure 8.
Attacked image with (a) JPEG quality 5, (b) JPEG quality 10, (c) JPEG quality 15, (d) Gaussian noise (variance = 0.0058), (e) impulse noise (normalized density of 0.015), (f) cropping, and (g) half sizing (followed by resizing back to the original size).

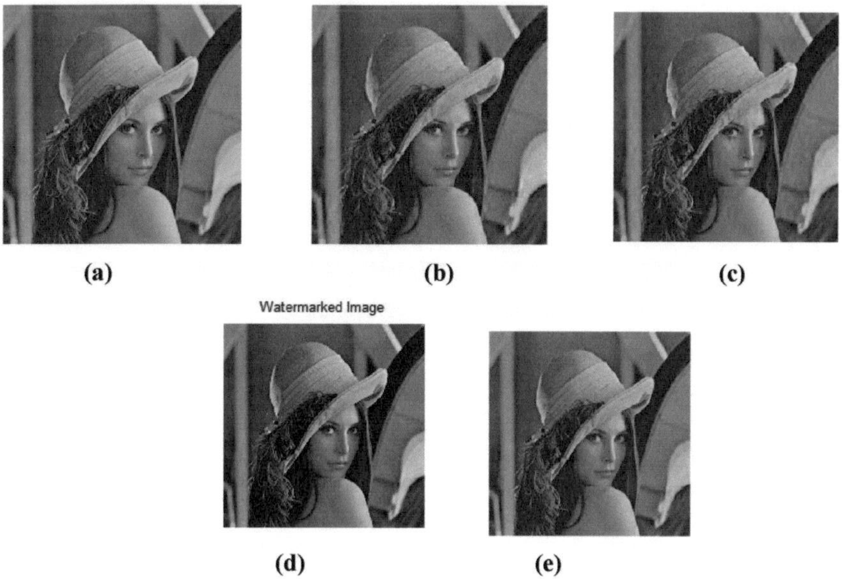

Figure 9.
(a) Original image. (b) Lena image marked using watermarking scheme of Dugad in the absence of attacks. (c) Lena image marked using watermarking scheme of Miyazaki in the absence of attacks. (d) Lena image marked using LSB scheme. (e) Lena image marked using the proposed watermarking method in the absence of attacks.

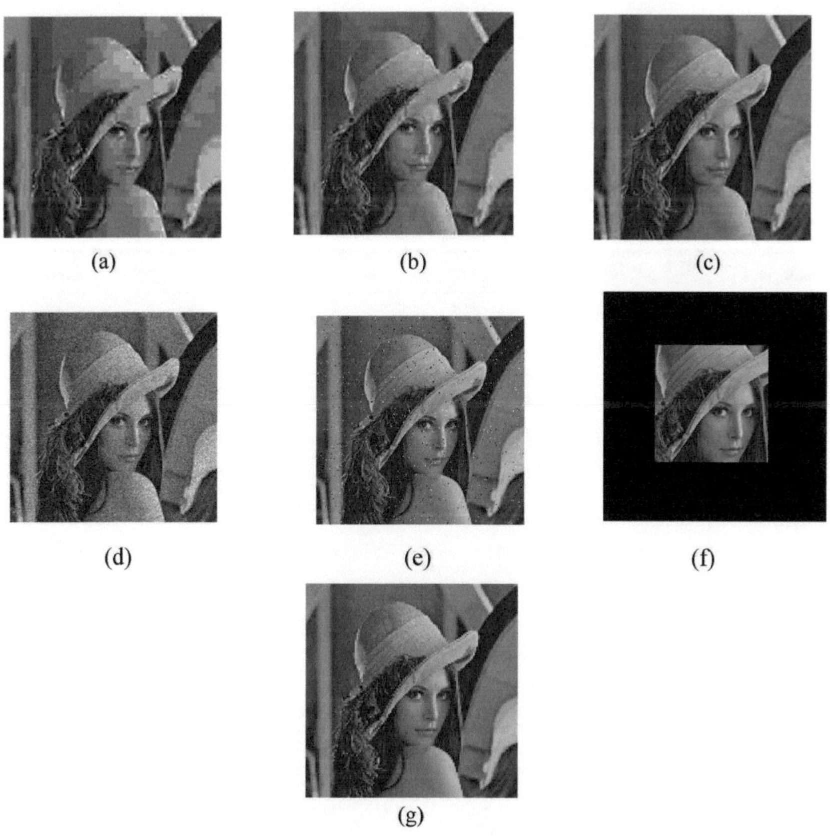

Figure 10.
Attacked image with (a) JPEG quality 5, (b) JPEG quality 10, (c) JPEG quality 15, (d) Gaussian noise (variance = 0.0058), (e) impulse noise (normalized density of 0.015), (f) cropping, and (g) half sizing (followed by resizing back to the original size).

Scheme	PSNR	NC
LSB scheme blind	51	1
Dugad scheme blind	40.09	0.45
Miyazaki scheme non-blind	44.62	1
Proposed scheme blind	45.36	1
Using t_1 = 90, t_2 = 200, and k = 0.1.		

Table 4.
Comparing the proposed method with the other two techs of Dugad, Miyazaki, and LSB (hat image).

NC				
Type of attacks	**Blind LSB scheme**	**Blind scheme in Dugad**	**Non-blind scheme of Miyazaki**	**Blind proposed method**
JPEG Q5	−0.0015	0.27	0.44	0.28
JPEG Q10	−0.0038	0.38	0.66	0.46
JPEG Q50	0.0015	0.45	1	0.88
Gaussian 0.006	0.0019	0.28	1	0.67
Gaussian 0.01	0.001	0.189	0.99	0.57
Gaussian 0.1	−8.1606e-007	0.01	0.46	0.05
Salt and pepper 0.015	−7.5838e-004	0.42	0.79	0.45
Salt and pepper 0.15	−0.0053	0.024	0.54	0.12
Salt and pepper 0.5	−0.0012	0.003	0.14	0.03
Cropping	−1.5895e-005	0.20	0.32	0.39
Half sizing	−0.0012	0.25	0.96	0.49
Subsample 0.7	−9.3287e-004	0.31	0.97	0.76
Subsample 0.4	−9.2566e-005	0.26	0.85	0.47

Table 5.
Comparing NC value for the proposed method with the methods of Dugad, Miyazaki, and LSB using hat image.

	NC	WM length in	WM length out
No attacks	1	367	367
JPEG Q5	0.14	367	203
JPEG Q10	0.48	367	271
JPEG Q15	0.85	367	319
Gaussian (0.006)	0.54	367	250
Salt and pepper (0.015)	0.79	367	293
Cropping	0.48	367	78
Half sizing	0.39	367	222

Table 6.
Results for the proposed scheme (hat image) with t_1 = 90, t_2 = 200, X_1 = 20 and X_2 = 10.

Scheme	PSNR	NC
LSB scheme blind	50.86	1
Dugad scheme blind	37.42	0.36
Miyazaki scheme non-blind	39.27	1
Proposed scheme blind	45.29	1
Using t_1 = 120, t_2 = 200, and k = 0.1.		

Table 7.
Comparing the proposed method with the other two methods of Dugad, Miyazaki, and LSB (Lena image).

NC				
Type of attacks	**Blind LSB scheme**	**Blind scheme of Dugad**	**Non-blind scheme of Miyazaki**	**Blind proposed method**
JPEG Q5	−0.0083	0.15	0.5	0.14
JPEG Q10	−0.0024	0.19	0.88	0.39
JPEG Q50	−0.0014	0.24	0.98	0.83
Gaussian 0.006	0.0038	0.24	0.9	0.32
Gaussian 0.01	0.0013	0.16	0.76	0.25
Gaussian 0.1	4.5235e-004	0.007	0.28	0.04
Salt and pepper 0.015	0.0016	0.18	0.96	0.6
Salt and pepper 0.15	8.2995e-004	0.03	0.35	0.53
Salt and pepper 0.5	5.2785e-004	0.005	0.059	0.47
Cropping	−0.0054	0.18	0.96	0.62
Half sizing	0.42	0.18	0.76	0.49
Subsample 0.7	0.56	0.25	0.83	0.67
Subsample 0.4	0.31	0.18	0.7	0.36

Table 8.
Comparing NC value for the proposed method with the methods of Dugad, Miyazaki, and LSB scheme using Lena image.

	NC	WM length in	WM length out
No attacks	1	129	129
JPEG Q5	0.14	129	57
JPEG Q10	0.39	129	74
JPEG Q15	0.83	129	102
Gaussian 0.006	0.32	129	66
Salt and pepper 0.015	0.6	129	41
Cropping	0.62	129	85

Table 9.
Results for the proposed scheme (Lena image) (with t_1 = 120, t_2 = 200, X_1 = 20 and X_2 = 10).

However in **Table 6**, the "cropping" attack poses a problem in that only 78 out of a possible 367 watermark bits were used by the detector, thus decreasing the reliability of the scheme. The scheme is not robust to JPEG quality 5 attacks. Thus, while surviving the same attacks as the Dugad scheme, the new scheme does not degrade the watermarked image to the same extent. From **Table 4**, PSNR value is 45.36 dB.

Table 7 presents the PSNR and NC for the proposed method and the other two methods using Lena image. It is seen that our method does not degrade the watermarked image to the same extent as the other two methods. **Table 8** represents the NC for the attacked watermarked images in our proposed method and the other existing methods.

Figure 11.
Watermarked image using HEAD method with DWT with and without attacks for Lena image. (a) Watermarked image PSNR = 51.7 dB without attacks. (b) Attacked image with Gaussian noise with variance = 0.006. (c) Cropped image. (d) Rotated image with 3°. (e) Resized image from 256 to 128–256. (f) Blurred image with 3 × 3 LPF.

Figure 12.
Watermarked image using the HEAD-based DWT method with and without attacks for hat image. (a) Watermarked image PSNR = 49.4 dB without attacks. (b) Attacked image with Gaussian noise with variance = 0.006. (c) Cropped image. (d) Rotated image with 3°. (e) Resized image from 256 to 128–256. (f) Blurred image with 3 × 3 LPF.

	Proposed HEAD watermarking method for Lena image, t1 = 182 and t2 = 268	Proposed HEAD watermarking method for Hat image, t1 = 236 and t2 = 333
No attacks	1	1
Gaussian 0.006	0.75	0.85
Gaussian 0.01	0.55	0.53
Gaussian 0.1	0.14	0.16
Cropping	1	1
Rotation	0.3	0.076
Blurring	0.6	0.43
Resizing 0.5	0.42	0.5

Table 10.
Correlation values for our scheme of Lena and hat images.

However in **Table 9**, the "salt-and-pepper noise" attack poses a problem in that only 41 out of a possible 129 watermark bits were used by the detector, thus decreasing the reliability of the scheme.

8.1 Simulation results of HEAD quantization method

We simulate the watermarking schemes on Lena and Hat images. Results are shown in **Figures 11** and **12**, respectively. The numerical evaluation metrics for all schemes in the absence and presence of attacks are tabulated in **Table 10**. From the table we notice that the proposed watermarking scheme achieves the lowest distortion in the watermarked image in the absence of attacks, and we find that the proposed method using wavelet gives the image with fidelity better than the other existing methods and the table gives the correlation under the presence of attacks; we notice also that a percentage of around 50% of the input watermark bits can be extracted in the proposed scheme with most of the attacks.

We find that we can detect watermark at the presence of blurring, Gaussian noise, cropping, and resizing attack; in the case of rotation attack, detection of watermark is difficult.

9. Conclusions

With this proposed method, blindness, detectability, robustness against attacks, and high watermarked image quality is maintained. Although the robustness of this new scheme is not quite as strong as that presented by Miyazaki method, this can be attributed to its blind nature compared to the semi-blind nature of the Miyazaki method. In LSB method, the attacks like addition of noise with any value or compression of the image using JPEG destroy the embedded watermark, and we cannot detect or extract the watermark at all, although the watermark was recovered perfectly in the ideal case.

Also the watermark may be removed without any effect done on the watermarked image. A blind DWT-based image watermarking schemes depend on the HEAD quantization of coefficients to embed meaningful information in the image. Experimental results have shown the superiority of the proposed schemes from the host image quality point of view, robustness, and the blindness point of view.

Author details

Abeer D. Algarni[1] and Hanaa A. Abdallah[1,2]*

1 Department of Information Technology, College of Computer and Information Sciences, Princess Nourah Bint Abdulrahman University, KSA

2 Electronics and Communications Department, Faculty of Engineering, Zagazig University, Egypt

*Address all correspondence to: haabdullah@pnu.edu.sa

References

[1] Zhu W, Xiong Z, Zhang Y. Multiresolution watermarking for images and video. IEEE Transactions on Circuits and Systems for Video Technology. 1999;**9**(4):545-550

[2] Xia XG, Boncelet CG, Arce GR. A multiresolution watermark for digital images. In: Proceedings of the IEEE International Conference on Image Processing (ICIP 1997); Vol. 1; October 1997. pp. 548-551

[3] Dugad K, Ratakonda R, Ahuja N. A new wavelet-based scheme for watermarking images. In: Proceedings of International Conference on Image Processing (ICIP 1998); Vol. 2; Chicago, IL; October 4–7, 1998. pp. 419-423

[4] Tao P, Eskicioglu AM. A robust multiple watermarking scheme in the DWT domain. In: Optics East 2004 Symposium, Internet Multimedia Management Systems V Conference; Philadelphia, PA; October 25–28, 2004. pp. 133-144

[5] Cox IJ, Miller ML, Bloom JA. Watermarking applications and their properties. In: Proceedings of the IEEE International Conference on Information Technology: Coding and Computing; 2000. pp. 6-10

[6] Tsai MJ, Yu KY, Chen YZ. Joint wavelet and spatial transformation for digital watermarking. IEEE Transactions on Consumer Electronics. 2000;**46**(1): 241-245

[7] Barni M, Bartolini F, Cappellini V, Piva A. A DCT-domain system for robust image watermarking. Signal Processing. Special Issue on Copyright Protection and Control. 1998;**66**(3): 357-372

[8] Potdar VM, Chang E. A Quantization Based Robust Image Watermarking Algorithm in Wavelet Domain. Perth, Western Australia: School of Information Systems, Curtin University of Technology. ICT_2005

[9] Miyazaki A, Yamamoto A, Katsura T. A digital watermarking technique based on the wavelet transform and its robustness on image compression and transformation. IEICE Transactions. Special Section on Cryptography and Information Security. 1999;**82**(1):2-10

[10] Zolghadrasli A, Rezazadeh S. Evaluation of spread spectrum watermarking schemes in the wavelet domain using HVS characteristics. The International Journal of Information Science and Technology. 2007;**5**(2): 123-139

[11] El-said SA, Hussein KFA, Fouad MM. Adaptive lossy image compression technique. In: Electrical and Computer Systems Engineering Conference (ECSE'10); 2010

Section 3

Data Mining in Cyberspace

Chapter 8

Text Mining to Facilitate Domain Knowledge Discovery

Chengbin Wang and Xiaogang Ma

Abstract

The high-precision observation and measurement techniques have accelerated the rapid development of geoscience research in the past decades and have produced large amounts of research outputs. Many findings and discoveries were recorded in the geological literature, which is regarded as unstructured data. For these data, traditional research methods have limited functions for integrating and mining them to make knowledge discovery. Text mining based on natural language processing (NLP) provides the necessary method and technology to analyze unstructured geological literature. In this book chapter, we will review the latest researches of text mining in the domain of geoscience and present results from a few case studies. The research includes three major parts: (1) structuralization of geological literature, (2) information extraction and visualization for geological literature, and (3) geological text mining to assist database construction and knowledge discovery.

Keywords: text mining, word segmentation, geological literature, visualization, knowledge discovery

1. Introduction

Geoscience is a knowledge-intensive discipline. It has not only domain-specific terminology but also a deep intersection with mathematics, chemistry, and physics, which form a series of distinctive subdisciplines, such as geophysics, geomathematics, geochemistry, paleobiology, and more [1–3]. Thanks to the rapid development of detection techniques in the micro- and macroscales in the past decades, both the volume and quality of geoscience data have been improved greatly. A feature of detection-based research is using the extrapolation method to explore the Earth. For instance, geochemists use local geochemical data to invert the process of Earth evolution and geodynamics [4, 5]. The diverse big data and improved computer software and hardware enable an opportunity to understand the evolution of Earth system using simulation and data mining methods [6].

Many geoscience research outputs are recorded in the form of literature, making text data an integral part of geoscience big data [7]. Important information and knowledge are recorded in unstructured textural form and thus hidden in the geological literature. Nowadays, the advanced Web technologies promote the publication process of academical literature and accelerate literature exchange globally. Researchers can easily assemble publications of focused topics. In this regard, geological literature has become a big "mineral resource" for data mining and provides tremendous opportunities for new knowledge discovery. In recent years, the open data initiative has promoted government agencies, scientific organizations, and academic publishers to provide literature archives for nonprofit reuse; some are even open and free. For instance, the US Geological Survey (USGS) and China Geological Survey (CGS) have published outputs of geological survey investigation online [8, 9]. Elsevier and Springer have provided application programming interfaces (API) for developer and scientists to access metadata, full text, and conduct text mining [10, 11]. We anticipate that more geological literature will be made available by publishers, government agencies, research organizations, and individual scientists in the coming years.

In a recent review article [12], Gil and other scholars proposed a research agenda of intelligent systems that will result in fundamental new capabilities for understanding the Earth system. Automated information extraction and integration from published literature is listed as a key research direction in the agenda. Domain-specific text mining can be regarded as a topic in interdisciplinary fields, such as geoinformatics, ecoinformatics, and bioinformatics. Conventionally, text mining is a research topic in computer science. The new development in interpreted programming language and the wide-spreading open-source packages and libraries enable scholars in various disciplines to quickly learn the latest algorithms and apply them to their domain-specific researches. There are many widely used open and free libraries in text mining, such as TensorFlow [13], DeepDive [14], Caffe [15], CNTK [16], and MXnet [17]. Even if a researcher has only the basic skills in programming, he or she will be able to make a deep research using these libraries.

Text mining contains the following major steps: data collection and preprocessing, identification of entities and their links, and knowledge representation. Data collection can take place in many forms. For example, one can require permission to get data from a database or publisher and can also retrieve data form the Web by a data extractor. The obtained data from different sources may be recoded in diverse formats, such as text files and scanned images. It is necessary to transform the data into an organized, computer-readable format. For instance, we can use the optical character recognition (OCR) to identify characters and words from the scanned images of a book or paper. After the preprocessing, the next step is to analyze the information and meaning of the text data. In the early stage, many researchers have tried to use automatic text summarization to extract a concise and informative abstract that covers the key information of a text document [18–20]. Nevertheless, due to the limitation of poor readability, automatic text summarization has yet to achieve satisfactory results.

Knowledge graphs, as proposed by Google, are semantic networks with directed graph structure, which have provided new ideas to extract and represent the text information. The words representing the major entities and relationships carry the key information in a document. Therefore, a text document can be represented by a knowledge graph to show a list of entities and their relationships. The structured knowledge graph is a specific data base and can be further analyzed and visualized by graph methods. Every entity is regarded as a graph node, and the relationship between two nodes is represented as an edge. The graph visualizes the nodes and edges to represent the implicit information network of a document. In recent years, many open knowledge graphs have been constructed based on text information, such as Google knowledge vault [21], DBpedia [22], Freebase [23], YAGO [24], Wikidata [25], OpenIE [26], and NELL [27]. These knowledge graphs devote to acquire entities and their links for various topics during the construction. In contrast, some domain-specific knowledge graphs only focus on one or a few topics. For instance, the MusicBrainz [28], UniProtKB [29], and GeoName [30] are knowledge graphs in the music, biology, and geography fields, respectively. The recent development of NLP and semantic technologies also provide new methods and tools for building knowledge graphs [14, 31, 32].

In this chapter, we will review the development of text mining in the domain of geoscience in recent years and present results of a few case studies. Comparing with other disciplines, the domain of geoscience still has limited applications of NLP and text mining. We hope the presented work will be of interested to the text mining community, and we anticipate more innovative text mining applications will appear in geoscience and other disciplines in the near future.

2. Structuralization of geological literature

Text data are usually consisted of sentences written by authors with personal understandings and opinions. Compared to metadata, text data are characterized by ambiguity, polysemy, and irregular input in the natural language. It is difficult for computers to read and understand. It is necessary to segment a piece of text into semantic word sequences for further computer processing. English and other Latin languages have relatively simple morphology, especially inflectional morphology, and are segmented by spaces between words naturally. For those languages, it is often possible to ignore the word segmentation task entirely. In contrast, there is no space between words in a few other languages, such as Chinese. It is difficult for a computer to identify the boundary of a meaningful word or phrase in Chinese [33, 34]. The methods of Chinese word segmentation were classified into dictionary-based, statistically based, and hybrid approaches [33]. The statistically based methods include machine learning and deep learning methods, such as hidden Markov model (HMM), maximum entropy Markov model (MEMM), conditional random fields (CRF), and long short-term memory (LSTM).

From another perspective, the methods of word segmentation can be divided into generic and specific domain methods according the usage scenarios. In the generic domain, because of the shortcomings of word segmentation rules, some new words, especially the professional terms, will be regarded as out-of-vocabulary and cannot be identified correctly. Geology, as a knowledge-intensive discipline, has a systematic domain-specific terminology. Most of geologic terms are not familiar with the public. Geological literature including the geological terms has their unique characteristics. For instance, the geological literature is always organized according to some fixed format and contains lots of professional geologic terms that only people with a background knowledge can read and understand. The geological literature is dominated by descriptive sentences and has little ambiguity in information expression. In geological literature written in Chinese, it is also featured by mixed writing of Chinese and English terms as well as compound terms consisted of multiple geological terms [2, 7]. The text data in the natural language are sequence data; the word usage and combination are only influenced by the context. Based on the characteristics of text data, machine learning method (e.g., CRF) and deep learning method of neural network (e.g., neural network (CNN), LSTM) have been introduced to segment geological literature in Chinese in recent 2 years with successful results [7, 34–36].

2.1 Conditional random fields

For a random vector (e.g., in NLP), the joint probability is a high-dimensional distribution, which oversteps the processing power of an ordinary computer and is difficult to monitor during data processing. To reduce the data size, the high-dimensional distribution is divided into a series of production of conditional probability based on the independence hypothesis [37]. The probabilistic graphical model is a graph to describe independence relationship between multivariate in a high-dimensional probabilistic model, thus to reduce computer load. The probabilistic graphical model includes both directed and undirected models. The directed graphical model indicates there is a causation relationship between the variables, such as Bayesian networks. The variables in undirected graphical model have dependency with each other, such as Markov networks and CRF, which is different from the causation relationships.

CRF model is a discriminative graph model, while HMM is a generative graph model. The role of CRF model is to create the discriminant boundaries similar to the support vector machine model, which has a wide usage in the fields of NLP and bioinformatics. Compared with HMM and the maximum entropy model (MEM), the CRF model improves the accuracy and addresses the drawback of label bias [38, 39]. Text data are unstructured sequence data. The structuralization of geological text is a process of word segmentation or named entity recognition (NER), which divides the geological text into a series of semantic words. For natural language, the text is only influenced by the context, which is consistent with the assumption condition of the CRF model. The assumption condition is that multivariable obeys the Markov property. In other word, the label of part of speech at *n* position in NLP only has relationship with the word or character at *n-1* position. From the point of view of the graph model, *Yv* is a subset of *V* nodes set in the graph $G = (V, E)$; the following equation is established:

$$p(Y_v|X, Y_w, w \neq v) = p(Y_v|X, Y_w, w \sim v) \tag{1}$$

where $\mathbf{X} = \{X_1, X_2, ..., X_n\}$ is the words or characters of text data in NLP, $w \sim v$ denotes neighbor nodes of node v in the graph, and Y is the label set of part-of-speech $\{B, E, M, S\}$. For NLP, the graphical structure is chain-structured (**Figure 1**) [14–16].

According the factorization of joint probability distribution of undirected graph, the CRF model can be written as

$$p(\mathbf{Y}|\mathbf{X}) = \frac{1}{Z(\mathbf{X})}\prod_i e^{\sum_k \lambda_k f_k(\mathbf{X}, Y_{i-1}, Y_i, i)} = \frac{1}{Z(\mathbf{X})} e^{\sum_i \sum_k \lambda_k f_k(\mathbf{X}, Y_{i-1}, Y_i, i)} \tag{2}$$

In which i is the node position, k denotes the sequence number of feature function, and λ_k is the weight parameter. In Eq. (2), the feature function can be expressed in Eq. (3), which contains information of transfer and status features.

$$f = \sum_i^T \sum_k^M \lambda_k f_k(\mathbf{X}, Y_{i-1}, Y_i, i) \tag{3}$$

In CRF-based word segmentation, Wang et al. [7] designed a two-step workflow to segment geological literature in Chinese. First, a hybrid corpus was created using

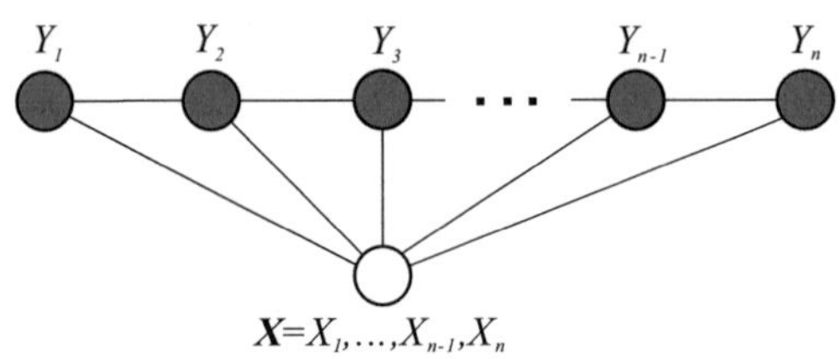

Figure 1.
A chain-structured CRF graph [7, 40].

dictionary matching and manual label methods on the basis of geological literature in CNKI, geology dictionary, TCCGMR (the terminologies and classification codes of geology and mineral resources), and a generic corpus of Peking University. Second, the segmentation rules were trained to build geological word segmentation model by the hybrid corpus, and then the trained model containing word segmentation rules was used to segment geological literature in Chinese. The workflow is shown in **Figure 2**.

In that study, a geology dictionary of 11,000 geological terms, the TCCGMR of 80,000 geological terms, and the generic corpus of Peking University were used to build the hybrid corpus. By this way, geological knowledge was introduced into the corpus to train the rules of word segmentation of geological literature. It is the most notable feature compared with other Chinese word segmentation machine. The three parameters of precision, recall, and F-scores were used to evaluate the performance of CRF-based word segmentation in that work. The result was showed in **Figure 3**. The hybrid corpus combining a generic corpus and a geological corpus has a better performance than either the generic corpus or the geological corpus alone. The precision of the hybrid training reaches 94.14%, which is 7.84% and 0.52% higher than that of CRF-PKU and CRF-GEO, respectively. The recall of hybrid corpus reaches 91.40%, which is 9.30% and 0.41% higher than that of CRF-PKU and CRF-GEO, respectively. The F-score of the hybrid corpus reaches 92.75%, which is 8.60% and 0.46% higher than that of CRF-PKU and CRF-GEO, respectively.

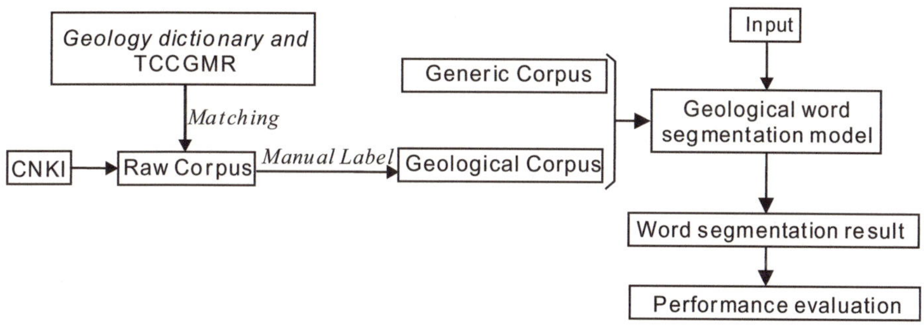

Figure 2.
Workflow of CRF-based word segmentation for geological literature in Chinese.

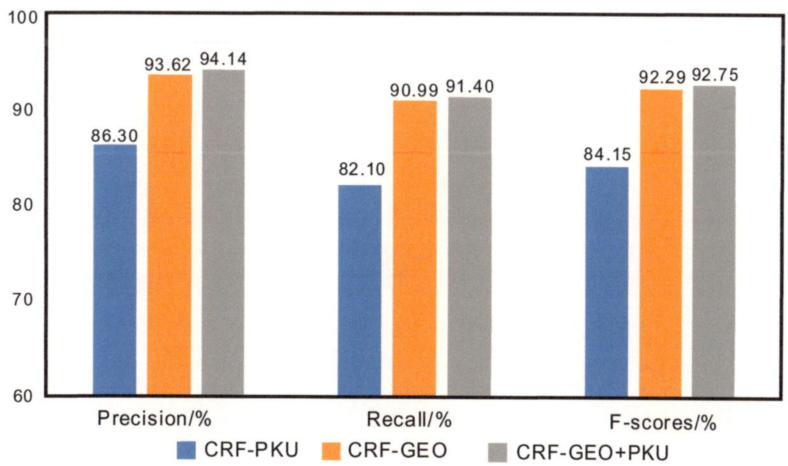

Figure 3.
Performances of the CRF model in different corpus. CRF-PKU, generic corpus of Peking University; CRF-GEO, geological corpus; CRF-GEO + PKU, the hybrid corpus combining generic and geological corpora.

2.2 Long short-term memory

The text data are consisted of a series of sequential words or characters, which can be regarded as a special data of time series and can be processed by the methods used in the time series analysis. Words or characters in text data are not completely independent but are connected to and influenced by the adjacent words or characters. In the model of neural network, it contains three basic compositions: input layer, hidden layer, and output layer. The layers of ordinary neural networks are linked with each other by weights. The nodes in a same layer are independent and have no link with each other. If the ordinary neural network methods are used to process text data, the semantic information of context will be missing. Recurrent neural network (RNN) has a short memory by nodes connecting in the hidden layer, which can receive information from self-cell and other cells. RNN has been used in the fields of NLP and automatic speech recognition [41, 42]. RNN model has the drawback of vanishing gradient problem, which means RNN model only obtains the information that is limited in the adjacent node position [43]. To address this challenge, the LSTM model designed input gate, output gate, and a forget gate to obtain information of far nodes and regulate the information flow between the cells [44] (**Figure 4**).

$$i_t = \sigma(W_i \cdot [h_{t-1}, x_t]) + b_i \tag{4}$$

$$f_t = \sigma(W_f \cdot [h_{t-1}, x_t]) + b_f \tag{5}$$

$$c_t = f_t c_{t-1} + i_t \tanh(W_c x_t + W_c h_{t-1} + b_c) \tag{6}$$

$$ot = \sigma(W_o \cdot [h_{t-1}, x_t]) + b_o \tag{7}$$

$$h_t = o_t{}^* \tanh(c_t) \tag{8}$$

in which i, f, c, and o denote input gate, forget gate, cell vector, and output gate, receptively. σ denotes the activation functions. W denotes weight matrices and bias vector parameters which need to be learned during the training.

Qiu et al. [36] proposed a geological literature segmenter based on the Bi-LSTM model. The segmenter was carried out by the following stages (more details can be seen in the reference article):

1. Corpus construction: The corpus from domain-generic and domain-specific texts is collected and constructed.

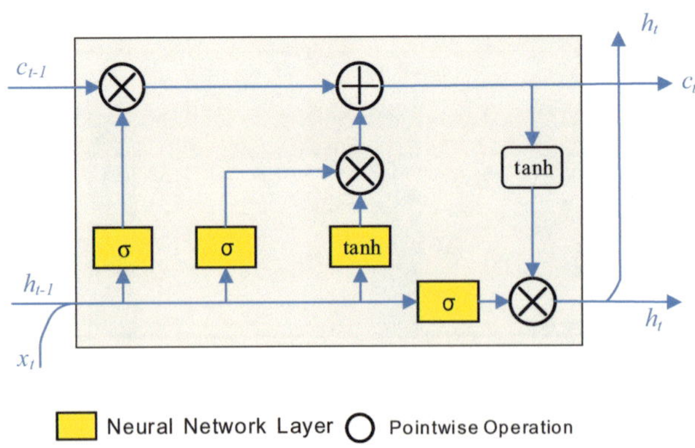

Figure 4.
The cell of LSTM [43, 45].

2. Words grouped: Each word is grouped based on frequency and a ranking algorithm.

3. Random extraction and combination: Each group of words in the previous step is extracted and joined together randomly.

4. Training: With the previous processing, sentences are formed via combination based on deep learning.

5. Testing and output: The resulting segmentation is post-processed and output.

In this research work, the significant highlight is that the training corpus is random. The segmentation rule was learned from the words and their corresponding sequences of the training corpus. The training corpus did not have any manual label information. The precision, recall, and F-scores reach 86.1%, 87.1%, and 86.6%, respectively. Compared with Wang et al. [7], the performance of CRF-based method is better than the Bi-LSTM-based segmenter based on the performance reported in their papers. But the Bi-LSTM-based method has a strong ability of identifying new words. The rate of out-of-vocabulary word identification reached 71.1%.

3. Text information visualization and knowledge discovery

3.1 Information visualization of a single geological literature

The nodes of content word and their links are the carrier of literature information and knowledge. In a large open knowledge graph, the key information was stored in in a triple format. Moreover, the bigram is also widely used in the text information representation. Wang et al. [7] used the bigram graph to represent the single geological literature.

The visualization was built based on the "from," "to," and "weight" variables. The variables of "from" and "to" indicate the sequence of content words in the content word corpus. In the content-word pairs, the former content word is defined as "from" variable, and the latter is defined as "to" variable. Their weights were defined by the co-occurrence frequency of content-word pairs. The bigram graph was used to visualize the nodes of content words and their links.

In geological exploration, the anomaly information of geology, geochemical exploration, geophysical exploration, and remote sensing is important clues for mineral prospecting [46]. To state different anomaly information, literatures of geological exploration will have significant features in the term of word frequency. **Figure 5** shows the main information hidden in a single literature of geophysical exploration. In this visualization, geological terms (e.g., *aeromagnetic*, *gravity*, *magnetic*) and geophysical data processing terms (e.g., *inversion*, *horizontal gradient*, *information*) are all linked to the term *anomaly*. The visualization represents the hidden key knowledge in the geological literature.

3.2 Geological text mining for discovering ore prospecting clues

Geology research not only reveals the earth evolution and promotes our understanding of the Earth but also has a close relationship with the human society. One of the important roles of applied geology is to discover mineral deposits and provide raw material for economic construction and development. In the long geological

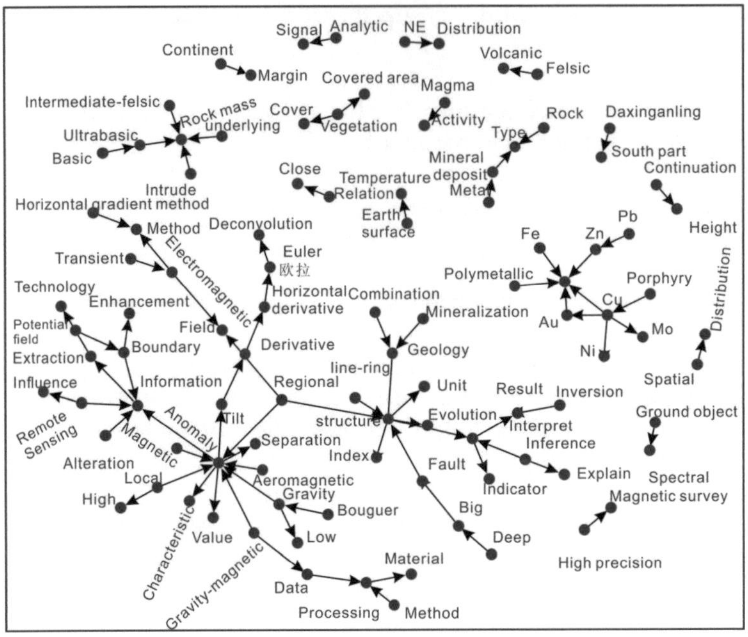

Figure 5.
Bigram graph of content words in the whole literature represents the key information in the geological report (n > 10).

history, mineral deposits were formed with large-scale geological events and were buried in the depth of Earth crust. If the mineral deposits were not broken down under the erosion of weathering after mineralization, there are ways to discover them. In the earlier days, geologists discovered mineral deposits by identifying the rock outcrops associated with mineralization. Then, along with many technological developments, the geochemical exploration, geophysical exploration, and remote sensing were also used to improve the result of mineral prospecting and mineral exploration. In recent years, GIS-based and three-dimensional mineral prospect mapping has been used in mineral exploration. Through those technologies, multisource anomalies, such as geochemical anomalies, geophysical anomalies, geological anomalies, and remote sensing anomalies, can be determined.

The anomaly information is usually derived from structured numeric data. The structured numeric data are only one part of geological big data. The majority of geological big data are unstructured, such as text and image. Previous mineral exploration mainly depends on derived information from the structured numeric data. Some important information related to the mineral prospecting and exploration is hidden in the unstructured text, such as host rock, alteration types, geological setting, ore-controlled factors, geochemical and geophysical anomaly patterns, and location. The favorable information extraction and identification from geological literature are a big challenge for conventional research methods. The NLP-based text mining provides a chance to address this challenge.

Li et al. [35] used the CNN method to classify geological text data into four categories (geology, geophysics, geochemistry, and remote sensing) on three scales (word, sentence, and paragraph). Their work extended the work on Chinese word segmentation and text preprocessing to the domain of mineral exploration. These four categories represent four types of mineral exploration information. Compared with word and paragraph scales, the sentence scale has the best performance. In their work, the *precision*, *recall* and *F-scores* of text classification reach 93.68%,

93.50%, and 92.68%, respectively. Then a co-occurrence matrix was utilized to extract content words and their relationships as nodes and links from the classification result and to visualize the information in a knowledge graph. By this way, four categories of favorable information for mineral prospecting and exploration were expressed in a bigram graph and a chord graph.

3.3 Geological text mining to assist database construction and knowledge discovery

The microfossil at 4280 million years old found in Quebec, Canada, may be the oldest fossil as so far [47]. In the Earth's history, biological evolution has a close corresponding with the geological evolution. The existence of biology depends on specific physical and chemical conditions, such as oxygen content and temperature. In other words, different biotypes and biocenoses indicate the conditions of different earth environments. The fossils were formed along with the sedimentary environment and are the footprint left by the biosphere. Each fossil records some biological information, such as biological morphology and living environment. Paleontologists always study the fossils to explore the earth environment evolution. A single fossil cannot indicate biological and geological evolution. The conclusions of such evolution are based on a series of comparative studies of fossils in different geological times and settings.

The Paleobiology Database (PBDB; http://paleobiodb.org) contains systematic and detailed fossil information, which make it a necessary infrastructure for fossil comparative researches. The PBDB is one of the most successful fossil databases, which was founded nearly two decades ago. Now it has become an open and active community for different research agendas. In the initial stage, the fossil records in the PBDB were from original fieldworks and extracted from published literature manually. As the rapid development of digital publication, the manual data entry for fossil information became tedious and less efficient and was not able to deal with the massive amounts of new and legacy publications. To address this challenge, PaleoDeepDive [46], a machine reading and learning system, was developed to extract fossil information from literature. This system uses the factor graph and NLP technologies to identify fossil entities and their semantic relationships. The extracted results were stored in the form of triples inside a knowledge base. Compared with the manual fossil data entry, the output of PaleoDeepDive has an obvious advantage in terms of quantity. Moreover, the change trend (e.g., taxonomic diversity and genus-level turnover) has a high corresponding relationship with the manual data entry [48]. The extracted fossil records have been used to update the PBDB. Now, the PBDB is not just a paleobiology database, it also provides WebGIS-based interface for fossil information retrieval and query. It also provides R library, API, and a mobile APP for researchers and the general public to use. Based on the PDBD, a series of high-quality research papers have been published to improve our understanding about the Earth. For instance, Peters et al. [49] analyzed the rise and fall of stromatolites in North America and divided the marine environment into three phases based the change of stromatolites.

The application of GeoDeepDive is still ongoing. Macrostrat (https://macrostrat.org/), a collaborative platform for geological data exploration and integration, was constructed based on the results that GeoDeepDive extracted from massive amounts of scientific literature. By April 2018, Macrostrat has contained 33,903 properties of geological units distributed across 1474 regions in North and South America, the Caribbean, New Zealand, and the deep sea, more than 180,000 geochemical and outcrop-derived measurements, all the fossil records in PBDB, and more than 2.3 million bedrock geologic map units from over 200 map sources [50].

4. Conclusion

In this chapter, we reviewed the latest developments of NLP techniques in the domain of geoscience to accelerate knowledge discovery from geological literature and deepen our understanding about the Earth. From the review, it was concluded that the researches of text mining in geoscience are still in the early stage. Most current researches focus on the literature structuralization and simple information extraction at a single document scale. The information integration and knowledge discovery from the big data of geological literature require further work and will lead to a lot of innovative research topics and applications.

Acknowledgements

The authors thank Prof. Abdelkrim El Mouatasim and Dr. Kristina Kardum for their organization for this book and their assistance. This work was supported by the National Key R&D Program of China (2017YFC0601500, 2017YFC0601504) and Fundamental Research Funds for the Central Universities, China University of Geosciences (Wuhan) (No. 162301182382).

Author details

Chengbin Wang[1*] and Xiaogang Ma[2]

1 School of Earth Resources, China University of Geosciences, Wuhan, China

2 Department of Computer Science, University of Idaho, Moscow, ID, USA

*Address all correspondence to: wangchb@cug.edu.cn

References

[1] Wang C, Ma X, Chen J. Ontology-driven data integration and visualization for exploring regional geologic time and paleontological information. Computers and Geosciences. 2018:12-19. DOI: 10.1016/j.cageo.2018.03.004

[2] Wang C, Ma X, Chen J. The application of data pre-processing technology in the geoscience big data. Acta Petrologica Sinica. 2018;**34**(2): 303-313. (in Chinese with English abstract)

[3] Ma X. Data science for geoscience: Leveraging mathematical geosciences with semantics and open data. In: Daya Sagar B, Cheng Q, Agterberg F, editors. Handbook of Mathematical Geosciences. Cham: Springer; 2018. pp. 687-702. DOI: 10.1007/978-3-319-78999-6_34

[4] Balch WM. Calcium carbonate measurements in the surface global ocean based on moderate-resolution imaging spectroradiometer data. Journal of Geophysical Research. 2005;**110** (C07001):1-21. DOI: 10.1029/2004jc002560

[5] Liu Y, Gao S, Hu Z, Gao C, Zong K, Wang D. Continental and oceanic crust recycling-induced melt–peridotite interactions in the trans-North China Orogen: U–Pb dating, Hf isotopes and trace elements in zircons from mantle xenoliths. Journal of Petrology. 2010;**51** (1–2):537-571. DOI: 10.1093/petrology/egp082

[6] Guo H, Liu Z, Jiang H, Wang C, Liu J, Liang D. Big earth data: A new challenge and opportunity for digital Earth's development. International Journal of Digital Earth. 2016;**10**(1):1-12. DOI: 10.1080/17538947.2016.1264490

[7] Wang C, Ma X, Chen J, Chen J. Information extraction and knowledge graph construction from geoscience literature. Computers and Geosciences. 2018;**112**:112-120. DOI: 10.1016/j.cageo.2017.12.007

[8] USGS. Mineral Resources Data System (MRDS) [Internet]. Available from: https://mrdata.usgs.gov/mrds/

[9] CGS. GEOCLOUD 2.0 [Available from: http://geocloud.cgs.gov.cn]

[10] Elsevier. Elsevier Developers-Text Mining [Internet]. Available from: https://dev.elsevier.com/tecdoc_text_mining.html

[11] Springer. Text and Data Mining at Springer Nature [Internet]. Available from: https://www.springernature.com/gp/researchers/text-and-data-mining

[12] Gil Y, Hill M, Horel J, Hsu L, Kinter J, Knoblock C, et al. Intelligent systems for geosciences. Communications of the ACM. 2018;**62**(1):76-84. DOI: 10.1145/3192335

[13] Google. TensorFlow 1.12.0 [Internet]. Available from: https://github.com/tensorflow/tensorflow/releases/tag/v1.12.0

[14] Zhang C. DeepDive: a data management system for automatic knowledge base construction[thesis]. Madison: University of Wisconsin-Madison; 2015

[15] Jia Y, Shelhamer E. Caffe Tutorial [Internet]. Available from: http://caffe.berkeleyvision.org/tutorial/

[16] Microsoft. The Microsoft Cognitive Toolkit [Internet]. Available from: https://www.microsoft.com/en-us/cognitive-toolkit/

[17] Apache. MXNet A flexible and efficient library for deep learning [Internet]. Available from: https://mxnet.apache.org/

[18] Hu B, Chen Q, Zhu F. LCSTS: A Large Scale Chinese Short Text Summarization Dataset. arXiv preprint arXiv:150605865. 2015

[19] Luhn HP. The automatic creation of literature abstracts. IBM Journal of Research and Development. 1958;**2**(2): 159-165

[20] Nallapati R, Zhou B, Gulcehre C, Xiang B. Abstractive text summarization using sequence-to-sequence rnns and beyond. arXiv preprint arXiv:160206023. 2016

[21] Dong X, Gabrilovich E, Heitz G, Horn W, Lao N, Murphy K, et al. Knowledge vault: A web-scale approach to probabilistic knowledge fusion. In: Proceedings of the 20th ACM SIGKDD International Conference on Knowledge Discovery and Data Mining. ACM; 2014

[22] Lehmann J, Isele R, Jakob M, Jentzsch A, Kontokostas D, Mendes PN, et al. DBpedia–a large-scale, multilingual knowledge base extracted from Wikipedia. Semantic Web. 2015; **6**(2):167-195. DOI: 10.3233/SW-140134

[23] Bollacker K, Evans C, Paritosh P, Sturge T, Taylor J. Freebase: A collaboratively created graph database for structuring human knowledge. In: Proceedings of the 2008 ACM SIGMOD International Conference on Management of Data. AcM; 2008. DOI: 10.1145/1376616.1376746

[24] Suchanek FM, Kasneci G, Weikum G. Yago: A core of semantic knowledge. In: Proceedings of the 16th International Conference on World Wide Web. ACM; 2007. DOI: 10.1145/1242572.1242667

[25] Vrandečić D, Krötzsch MJCotA. Wikidata: A free collaborative knowledgebase. 2014;**57**(10):78-85. DOI: 10.1145/2629489

[26] Stanovsky G, Dagan I. Open IE as an intermediate structure for semantic tasks. In: Proceedings of the 53rd Annual Meeting of the Association for Computational Linguistics and the 7th International Joint Conference on Natural Language Processing (Volume 2: Short Papers). 2015

[27] Mitchell T, Cohen W, Hruschka E, Talukdar P, Yang B, Betteridge J, et al. Never-ending learning. Communications of the ACM. 2018;**61**(5):103-115. DOI: 10.1145/3191513

[28] Hemerly J. Making metadata: The case of MusicBrainz; 2011. DOI: 10.2139/ssrn.1982823

[29] Boutet E, Lieberherr D, Tognolli M, Schneider M, Bansal P, Bridge AJ, et al. UniProtKB/Swiss-Prot, the manually annotated section of the UniProt KnowledgeBase: How to use the entry view. Methods in Molecular Biology. 2016;**1374**

[30] Maltese V, Farazi F. A semantic schema for GeoNames [Internet]. 2013. Available form: http://eprints.biblio.unitn.it/4088/1/techRep004.pdf

[31] Tseng Y-H, Lee L-H, Lin S-Y, Liao B-S, Liu M-J, Chen H-H, et al., editors. Chinese open relation extraction for knowledge acquisition. In: Proceedings of the 14th Conference of the European Chapter of the Association for Computational Linguistics. Vol. 2: Short Papers. 2014

[32] Zheng X, Li S, Feng J, Lin M, Song H, Zhang S. FudanDNN: A Deep Learning Framework with Easy-to-use GUI [Internet]. Available from: https://github.com/FudanDNN/FudanDNN

[33] Gao JF, Li M, Wu A, Huang CN. Chinese word segmentation and named entity recognition: A pragmatic approach. Computational Linguistics. 2005;**31**(4):531-574. DOI: 10.1162/089120105775299177

[34] Huang L, Du YF, Chen GY. GeoSegmenter: A statistically learned Chinese word segmenter for the

geoscience domain. Computers and Geosciences. 2015;**76**:11-17. DOI: 10.1016/j.cageo.2014.11.005

[35] Li S, Chen J, Xiang J. Prospecting information extraction by text mining based on convolutional neural networks–a case study of the Lala copper deposit, China. IEEE Access. 2018;**6**:52286-52297. DOI: 10.1109/access.2018.2870203

[36] Qiu Q, Xie Z, Wu L, Li WJ. DGeoSegmenter: A dictionary-based Chinese word segmenter for the geoscience domain. Computers and Geosciences. 2018;**121**:1-11. DOI: 10.1016/j.cageo.2018.08.006

[37] Sutton C, McCallum A. An introduction to conditional random fields. Foundations and Trends® in Machine Learning. 2012;**4**(4):267-373

[38] Lafferty JD, McCallum A, Pereira FCN. Conditional random fields: Probabilistic models for segmenting and labeling sequence data. In: Proceedings of the Eighteenth International Conference on Machine Learning. 655813. Morgan Kaufmann Publishers Inc; 2001. pp. 282-289

[39] Pinto D, McCallum A, Wei X, Croft WB. Table extraction using conditional random fields. In: Proceedings of the 26th Annual International ACM SIGIR Conference on Research and Development in Information Retrieval; Toronto, Canada. 860479. ACM; 2003. pp. 235-242

[40] Wallach HM. Conditional Random Fields: An Introduction [Internet]. Available from: http://dirichlet.net/pdf/wallach04conditional.pdf

[41] Yin W, Kann K, Yu M, Schütze H. Comparative study of CNN and RNN for natural language processing. 2017. arXiv:1702.01923

[42] Graves A, Liwicki M, Fernández S, Bertolami R, Bunke H, Schmidhuber J, et al. A novel connectionist system for unconstrained handwriting recognition. IEEE Transactions on Pattern Analysis and Machine Intelligence. 2009;**31**(5): 855-868. DOI: 10.1109/TPAMI.2008.137

[43] Bengio Y, Simard P, FRasconi P. Learning long-term dependencies with gradient descent is difficult. IEEE transactions on neural networks. 1994;**5**(2):157-166

[44] Hochreiter S, Schmidhuber J. Long short-term memory. Neural Computation. 1997;**9**(8):1735-1780. DOI: 10.1162/neco.1997.9.8.1735

[45] Olah C. Understanding LSTM networks. [Internet]. 2015. Available form: http://colah. github. io/posts/2015-08-Understanding-LSTMs

[46] Wang C, Rao J, Chen J, Ouyang Y, Qi S, Li Q. Prospectivity mapping for "Zhuxi-type" copper-tungsten polymetallic deposits in the Jingdezhen region of Jiangxi Province, South China. Ore Geology Reviews. 2017;**89**:1-14. DOI: 10.1016/j.oregeorev.2017.05.022

[47] Dodd MS, Papineau D, Grenne T, Slack JF, Rittner M, Pirajno F, et al. Evidence for early life in Earth's oldest hydrothermal vent precipitates. Nature. 2017;**543**(7643):60. DOI: 10.1038/nature21377

[48] Peters SE, Zhang C, Livny M, Re C. A machine reading system for assembling synthetic paleontological databases. PLoS One. 2014;**9**(12): e113523. DOI: 10.1371/journal.pone.0113523

[49] Peters SE, Husson JM, Wilcots J. The rise and fall of stromatolites in shallow marine environments. Geology. 2017;**45**(6):487-490. DOI: 10.1130/G38931.1

[50] Peters SE, Husson JM, Czaplewski J. Macrostrat: A platform for geological data integration and deep-time earth crust research. Geochemistry, Geophysics, Geosystems. 2018;**19**(4): 1393-1409. DOI: 10.1029/2018GC007467

Chapter 9

Tagging and Tag Recommendation

Fabiano M. Belém, Jussara M. Almeida and Marcos A. Gonçalves

Abstract

Tagging has emerged as one of the best ways of associating metadata with objects (e.g., videos, texts) in Web 2.0 applications. Consisting of freely chosen keywords assigned to objects by users, tags represent a simpler, cheaper, and a more natural way of organizing content than a fixed taxonomy with a controlled vocabulary. Moreover, recent studies have demonstrated that among other textual features such as title, description, and user comments, tags are the most effective to support information retrieval (IR) services such as search, automatic classification, and content recommendation. In this context, tag recommendation services aim at assisting users in the tagging process, allowing users to select some of the recommended tags or to come up with new ones. Besides improving user experience, tag recommendation services potentially improve the quality of the generated tags, benefiting IR services that rely on tags as data sources. Besides the obvious benefit of improving the description of the objects, tag recommendation can be directly applied in IR services such as search and query expansion. In this chapter, we will provide the main concepts related to tagging systems, as well as an overview of tag recommendation techniques, dividing them into two stages of the tag recommendation process: (1) the candidate tag extraction and (2) the candidate tag ranking.

Keywords: tagging, folksonomies, Web 2.0, tag recommendation, keyword extraction, tag ranking

1. Introduction

Web 2.0 applications are characterized by the central role played by users in the creation and sharing of their own content. Tagging has become a common feature available in these applications, consisting in associating freely created tags (keywords) to objects (e.g., videos, images, texts). In comparison with a fixed taxonomy, tags are simpler, cheaper, and a more natural way of organizing content. In fact, taxonomies with a controlled vocabulary do not suit the increasing and evolving Web 2.0 environment [1].

Moreover, various studies have demonstrated that, among other textual features such as title, description, and user comments, tags are the most effective to support information retrieval (IR) services such as search [2], automatic classification [3], and content recommendation [4].

The tagging process can benefit a lot from a tag recommendation service. This type of service supports users in the selection of some of the recommended tags or in the creation of new ones. With that in mind, tag recommendation benefits are not limited to the improvement of the user experience: there is a high potential of improving the quality of the generated tags by, for example, reducing the amount of misspellings and nondescriptive keywords. Thus, the quality of the IR services that rely on tags as data sources can be indirectly improved by tag recommendation. Other examples of the benefits that tag recommendation can bring to IR services include the direct application of the recommended tags in search [5] and on query expansion [6]. In search, the recommended tags can be exploited to measure the similarity between queries and documents, improving the quality of the retrieved documents. Query expansion, in turn, aims at suggesting more specific and unambiguous queries to the user, which also allows the achievement of better search results. Further examples include researcher profile summarization [7] and search result summarization [8].

Tag recommendation brings specific challenges that other kinds of recommendation services do not: in the tag domain, we are interested not only in matching the interests of the target user but also in describing, summarizing, and organizing Web content. Thus, the design of tag recommenders demands specific solutions which greatly differ from methods proposed for item recommendation tasks in general. For instance, text mining, knowledge extraction, and semantics play a substantial role in the tag domain. In sum, the recommendation effectiveness affects not only user satisfaction but also the performance of various IR services that rely on tags as data source.

The goal of this chapter is to present the concepts of tagging systems and to provide an overview of tag recommendation techniques, explaining the two main steps of these methods: the candidate tag generation and the candidate tag ranking.

The rest of this chapter is organized as follows. In Section 2, we define tags, objects, folksonomies, and other basic concepts related to tagging systems. In Section 3, we state the tag recommendation problem, while we explain the main tag candidate extraction and ranking techniques in Sections 4 and 5, respectively.

2. Tags and Web 2.0 objects

A Web 2.0 *object* or *resource* (e.g., a textual document, audio image, or video) is defined as the main content of a Web 2.0 page. There are various sources of data related to this object, here referred to as its *features*, which we can classify as content features, textual features, user profile features, and social features.

Content features are attributes that can be extracted from the main content of the Web 2.0 object, such as the color histogram of an image. *Textual features*, in turn, comprise the self-contained textual blocks that are associated with an object, usually with a well-defined functionality, such as title, description, categories, tags, and user comments [3]. Note that these two sets of features may not be disjoint (e.g., when the main object is a textual document).

In particular, *tags* are keywords freely created by users and associated with objects. Tags are not necessarily unigrams (unless the application automatically splits them by whitespaces). Thus, tags may be composed by two or more words, sometimes separated by spaces, hyphenated, or joined.

Figure 1 illustrates a MovieLens page containing textual features assigned to an object (a movie, in this case).

User profile features include characteristics of the users who created or interacted with the content, while *social features* refer to interactions among users (e.g., explicit friendship links, subscriptions, "likes," etc.). The social connections among users may be explicitly represented by friendship links or implicitly indicated by subscriptions (connections established among users that show interests in one another's content), and endorsements (e.g., "likes"). **Figure 2** illustrates examples of these features.

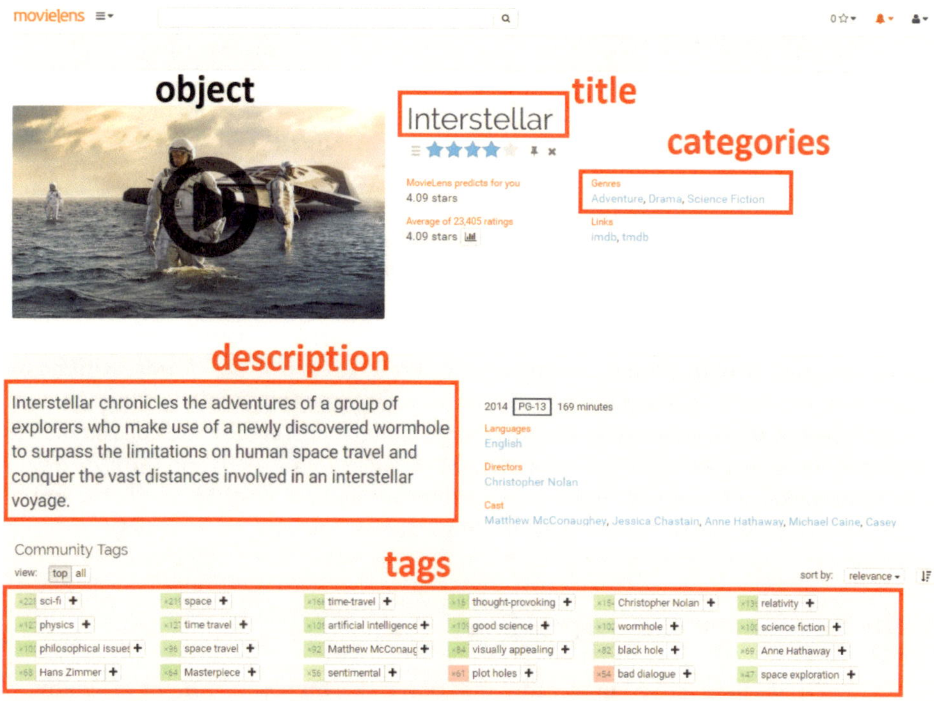

Figure 1.
A Web 2.0 page and some of its textual features.

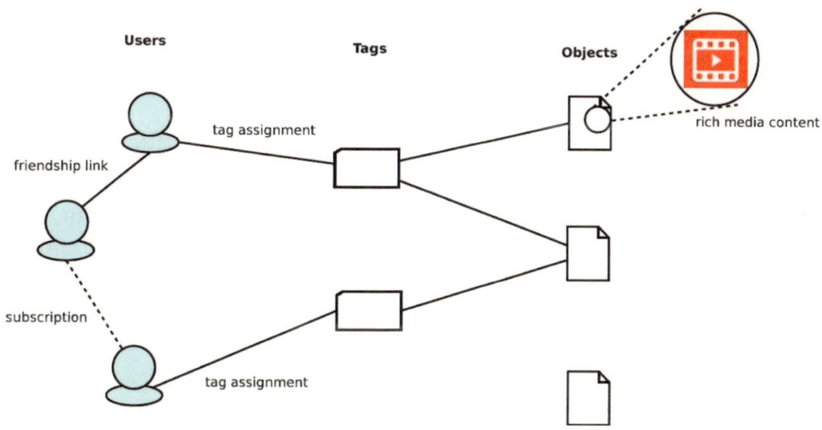

Figure 2.
Features commonly found in Web 2.0 pages. Friendship and subscription links are representative examples of social features. The set of tags a user assigned to objects in the applications is taken as one of the user profile features. Features extracted from the content of the main object (e.g., color histogram) are examples of content features [9].

The Web 2.0 tags, objects, and users form the basic structure of the *folksonomies*, which are defined as the categorization of objects using freely chosen keywords by users. Unlike a taxonomy, which provides a hierarchical categorization with well-defined classes, a folksonomy establishes categories (as tags) without imposing a hierarchical structure [10].

More formally, a folksonomy is defined as a relation $F = (U, T, O, P)$, where U, T, and O are finite sets composed by users, tags, and objects, respectively, and P, the set of postings, is a ternary relation between these elements, that is, $P \subseteq U \times T \times O$ [11].

Thus, each element $(u, t, o) \in P$ indicates that a user u associated a tag t to an object o (this is illustrated as the edges connecting users, tags, and objects in **Figure 2**). In [12], folksonomies are classified in *broad* and *narrow* folksonomies. A broad folksonomy occurs when multiple users can apply the same tag to an object, while a narrow folksonomy occurs when only one user (typically the target object's creator) can tag a given object.

Examples of broad folksonomies include the online radio station LastFM (http://www.last.fm/) and the publication sharing application Bibsonomy (http://www.bibsonomy.org). The photo sharing site Flickr (http://www.flickr.com/) is an example of narrow folksonomy. While both broad and narrow folksonomies have common goals, a broad folksonomy can be further exploited to rank tags by their popularity and visualize the most important tags by means of tag clouds, which also provide an easy way to navigate the tags, objects, and users of a folksonomy.

Examples of tagging datasets available online for experimentation include MovieLens and Bibsonomy snapshots (https://grouplens.org/datasets/movielens and http://www.kde.cs.uni-kassel.de/bibsonomy/dumps, respectively) and our LastFM, YouTube, and YahooVideo crawled data (https://figshare.com/articles/data_tar_gz/2067183).

3. The tag recommendation problem

As in [13], we define two tag recommendation tasks: the *object-centered* problem and the *personalized* problem. In the former, the goal is to generate and rank candidate tags according to their relevance to the target object, that is, the extent to which the tag is related to or describes the target object. *Object-centered tag recommendations,* which do not vary according to the target user, aim at improving tag quality and indirectly improving the effectiveness of information retrieval services, such as searching, classification, and item recommendation, which exploit tags as data sources.

On the other hand, *personalized* tag recommendation takes not only the target object but also the target user into account, aiming at suggesting tags that are relevant to both the target object and the user. Thus, personalized tag recommenders might provide different results for different users, which may better capture the user interests, profile, and background. According to [13], "in applications where multiple users can assign tags to the same object, such as Last.FM, a personalized tag recommender is not only useful for the individual user (e.g., for content organization) but also in a collective sense. This is because, jointly, the tags recommended to different users may provide a more complete description of the object, benefiting search and recommendation services."

In more formal terms, the tag recommendation tasks are defined in [13] as:

> "**Object-Centered Tag Recommendation.** *Given a set of input tags I_o associated with the target object o, generate a list of candidate tags C_o, sorted according to their relevance to object o, and recommend the k candidates in the top positions of C_o.*
>
> **Personalized Tag Recommendation.** *Given a set of input tags I_o, associated with the target object o, generate a list of candidate tags $C_{o,u}$, sorted according to their relevance to both user u and object o, and recommend the k candidates in the top positions of $C_{o,u}$."*

Note that possibly there are no tags available in the target object, that is, $I_o = \emptyset$. This is a variation of the "cold start" problem, a well-known problem in recommender systems generally defined as a scenario in which there is an insufficient amount of information about the target user or object, making it difficult to provide effective recommendations.

These definitions focus on relevance as the only objective to be maximized. However, other aspects of the problem, such as novelty and diversity, have been considered as important, in recommendation systems in general and also in the specific tag recommendation domain [14].

According to the traditional definition of relevance or accuracy, the relevance of each tag in a recommendation list is independent of the relevance of the other tags in the list. However, in the general recommendation context, given that a recommendation satisfied the user need, the usefulness of similar recommendations is arguable. This occurs in the tagging context when, for example, only synonyms or strongly similar words are provided as recommendations. To deal with these issues, concepts of novelty and diversity have been introduced.

In tag recommendation, the novelty of a tag has been defined from the perspective of its popularity in the application. In [14], tag novelty is calculated as the inverse of the frequency at which the tag is used in the collection. The rationale is that frequently used tags tend to be more "obvious" recommendations (if relevant), thus being of little use to improve the description of the target object. We note that, according to this definition, noisy terms such as typos may be considered highly novel. However, novelty and diversity must be considered jointly with relevance in order to provide effective tag recommendations. It is worth mentioning that this definition of novelty is closely related to tag *specificity* [15], since rare words tend to be more specific (less general). For example, the word "feline" is less specific than "cat" or "tiger," and thus it is expected that "feline" would be used to describe a larger number of objects than these more specific terms. Therefore, specificity can be interpreted as a statistical property of the term use, being estimated as an inverse function of the frequency of the tag in the collection [14].

The *diversity* of a list of recommended tags, in turn, can be interpreted as the *exhaustivity* of these tags, which is defined in [15] as the coverage they provide for the topics of the associated object. Two approaches to estimate diversity in tag recommendation have been proposed. The implicit approach exploits properties of the recommended items (tags in our case), estimating diversity as the average pairwise semantic dissimilarity between the top recommended tags. In this context, a list of synonyms or semantically related words presents low diversity [14]. The explicit diversification approach, on the other hand, exploits properties of the target of recommendations, such as a set of explicit topics (e.g., categories) related to the target object. The goal of the explicit diversifiers is to cover as many topics related to the target object as possible, and as early in the ranking as possible, minimizing redundancy, that is, focus on a single topic.

Table 1 summarizes the tag recommendation problem and its aspects.

	Personalized	**Object-centered**
Input	I_o: set of input tags associated with the target object o	
Target	Object o	Pair object-user $<o,u>$
Output: ranked list of candidate tags	C_o: sorted according to relevance and other aspects related to o	$C_{o,u}$: sorted according to relevance and other aspects related to the pair $<o,u>$
Other aspects		
Novelty/specificity: capacity of recommending more rare tags **Diversity/exhaustivity**: capacity of recommending tags related to the different topics of the target object or user		

Table 1.
Tag recommendation: problem statement.

4. Candidate tag generation

Tag recommendation methods can be divided in two steps: (1) the generation of a set of candidate tags and (2) the ranking of the candidate tags produced in step (1). In this section, we introduce the main techniques to tackle the first step, while in Section 5, we discuss methods to perform the second step.

The candidate tag generation depends on the data sources available in the target application. As summarized by [9], previous tag recommendation strategies have exploited as data sources: (1) the folksonomy (history of tag assignments); (2) textual features (other than tags), such as title, description, and user comments; (3) rich media content, that is, image, audio, or video; and (4) social features, such as friendship links in social networks and other interactions among users as illustrated in Section 2.

Based on these data sources, we can name three main groups of techniques to extract or generate candidate tags: (1) extraction of terms from the textual features associated with the target object, (2) tag co-occurrences with terms in these textual features (possibly including previously assigned tags) or other features (e.g., visual features for rich media content), and (3) tags extracted from neighbors, that is, objects that are similar to the target object or users that are similar to the target user. These three groups of techniques will be the subject of Sections 4.1–4.3, respectively.

4.1 Keyword extraction from texts

The simplest strategy to extract candidate tags from a given text is to consider each (whitespace) separated word as a candidate, after removing punctuation and other special characters. After this, a basic post-processing step is to remove *stop words* (i.e., words such as articles, prepositions, and conjunctions, which carry little semantics and thus are not adequate as keywords) from the list of generated candidates. Finally, corpus-oriented statistics of these individual words are evaluated to select the most promising candidates. These statistics are also exploited to rank candidate tags, and thus they will be discussed in Section 5.

However, this simple strategy is only capable of generating single words as tags, although it is common to use expressions containing two or more words (e.g., "information systems," "digital image processing") as tags. Thus, alternative keyword extraction techniques first generate all word n-grams obtained from a sliding window through the text, for n ranging from one to, let us say, three or four words. For example, for the following sentence:

"The tagging process can benefit a lot from a tag recommendation service."

A sliding window of size $n = 3$ would produce the following initial terms as keywords:

The tagging process – tagging process can – process can benefit – can benefit a – benefit a lot – a lot from – lot from a – from a tag – a tag recommendation – tag recommendation service.

To filter out meaningless or uninformative candidate tags such as "benefit a lot" or "from a tag," some authors, such as [7, 16] exploit a selection approach based on part-of-speech (PoS) labels, which captures the idea of keywords having a certain syntactic property. Besides that, this approach is based on empirical evidence obtained in training data. First, the most frequent PoS patterns of keywords that occur in a given training dataset are identified. For example, the three most frequent PoS patterns for keywords found in [15] are:

- ADJECTIVE + NOUN (singular or plural)
- NOUN + NOUN (both singular or plural)
- ADJECTIVE + NOUN (plural)

Thus, only sequences of words that match the top-x (let us say, x = 50) most frequent patterns are selected as candidate tags. For the aforementioned example and considering n = 2 and x = 3, the selected candidate tags would be "tagging process," "tag recommendation," and "recommendation service", all three of them matching the ADJECTIVE + NOUN pattern.

Unlike the PoS-based approach, which is a supervised, language-dependent approach that processes a training dataset, the *Rapid Automatic Keyword Extraction* (RAKE) [17] relies only on the target text to generate keywords, being known as a "document-oriented" approach, as opposed to the "corpus-oriented" methods. RAKE is based on the observation that keywords frequently contain multiple words but rarely contain standard punctuation or stop words. Instead of using an arbitrarily sized sliding window, RAKE splits the text using stop words and punctuation as delimiters. In our sentence example, "**The** tagging process **can** benefit **a lot from a** tag recommendation service," the stop words (in bold) would be discarded, generating the following candidate tags:

tagging process – benefit – tag recommendation service

After extracting candidate keywords, RAKE builds a graph of word co-occurrences, in which there is an edge between two words if they appeared in the same keyword. The score of each word w is calculated as $deg(w)/freq(w)$, where $deg(w)$ is the degree of w in the co-occurrence graph and $freq(w)$ is the number of occurrences of w in the text. The score of a given candidate keyword is defined as the sum of the scores of its containing words. Finally, in order to consider keywords that contain stop words (e.g., "set *of* natural numbers"), pairs of candidate keywords that appear in consecutive positions of the text at least twice are adjoined.

4.2 Tag co-occurrences

Another strong source of candidate tags is the history of tag assignments of the application (folksonomy). Tags that the target user frequently used in previous tagging events are good candidates to recommend for this user, especially in a personalized recommendation task. Still more interesting, we can exploit tag co-occurrences in these previous posts, recommending to an object o, with an initial set of tags I_o, tags that frequently co-occur with the tags in I_o in a training folksonomy dataset D, as performed by [2, 13, 14].

Tag co-occurrences are usually computed by exploiting association rules, which are employed in general to describe frequently co-occurring item sets. For tag recommendation, association rules assume the form $X \rightarrow y$, where X (the antecedent) is a set of tags and y (the consequent) is a candidate tag for recommendation. The main metrics that estimate the strength of an association rule are the *support*, defined as the number of co-occurrences of X and y in the training set, and the confidence, calculated as the conditional probability that y is assigned as a tag to an object given that all tags in X are also associated with it. Considering that the number of rules extracted from the training set can be very large and some of them may not be useful for recommendation, minimum support and confidence thresholds are used as lower bounds to select only the most important and/or reliable rules. This selection can improve both effectiveness and efficiency of the recommender.

To recommend tags for an object o, we select rules $X \rightarrow y$ in which X is a subset of I_o, the set of initial tags in o. For each term c appearing as consequent of any of the selected rules, we usually estimate its relevance as a tag for the object (and for the user in the personalized case), given the initial tag set I_o, as the sum of the confidences of all rules containing c. In the absence of an initial tag set, words occurring in other textual features of the target object, such as title and description, can be used as I_o, as performed by [18]. Another alternative is to compute co-occurrences between tags and visual features extracted from images or other rich media content associated with the target object [19].

4.3 Tags from neighbors

Another form of obtaining candidate tags that are external to the target object, besides exploiting co-occurrences, is extracting tags from the neighborhood of the target object o, that is, the set of most similar objects with relation to o. Similarly, we can generate candidate tags for a target user u from similar users or users that have some kind of connection in the application (e.g., explicit friendship links, endorsement links, etc.). The rationale is that similar objects or users are usually associated with similar tags.

Thus, the neighborhood-based tag generation approaches exploit a graph in which the nodes correspond to objects or users, and there is an edge between two objects (or two users) if they are similar (e.g., share tags or other words in common). Alternatively, visual features extracted from image and video objects can be used to estimate content similarity [19, 20], although they may face scalability issues and a larger semantic gap [20].

To identify similar objects or users, each object (or user) is usually modeled as a bag of terms (extracted from the textual features of the object or from the vocabulary of the users). These terms receive a TFIDF weight, and a similarity measure such as the cosine of these term vector representations is exploited to estimate the similarity between objects or users [21].

5. Candidate tag ranking

After generating a set of candidate tags, it is necessary to rank them, showing the most relevant tags first, in order to provide effective tag recommendations. Some tag candidate generation strategies already provide a measure to estimate the candidate tag relevance, such as the degree/frequency ratio in RAKE, as defined in Section 4.1. In this section, we first discuss various tag quality attributes that can be used to estimate tag relevance (Section 5.1), isolated or combined with other attributes. Then, we discuss methods that can automatically combine various attributes exploiting a learning-to-rank approach (Section 5.2).

5.1 Tag quality attributes

Tag quality attributes can be grouped into the following categories, based on the aspect they try to capture regarding the tag recommendation task [13]:

- *Tag co-occurrence attributes*: estimate how relevant a candidate tag c is given a set of input tags that often co-occur with c in the data collection.
- *Descriptive power attributes*: estimate how accurately a candidate tag describes the object's content based on statistics of the occurrence of the tag in the textual features of the target object.
- *Discriminative power attributes*: estimate the capability of a candidate to distinguish the target object from others.
- *Term predictability*: indicates the likelihood that a word can be predicted as a tag.
- *User interest attributes*: used for personalization, these attributes estimate the interest of a target user in certain tags.

5.1.1 *Tag co-occurrence attributes*

As mentioned in Section 4.2, tag recommenders select association rules in which antecedents are included in I_o, the set of tags already available in the target object, or terms that can be used as proxy for these initial tags. For each tag c appearing as consequent of any of the selected rules, the relevance of c as a tag for the object o, given the initial tag set I_o, can be estimated by sum, which sums up the confidences of all rules that point to c, i.e.:

$$Sum(c, I_o) = \sum_{X \subseteq I_o} confidence(X \rightarrow c), (X \rightarrow c) \in R, |X| \leq l, \tag{1}$$

where R is a set of association rules generated from the training set and l is the size limit for the association rules' antecedents, usually limited to 1 or 2 words, due to performance issues.

Sum was proposed by [2], which also proposed several other attributes related to tag co-occurrences. For example, *Vote* (c, I_o) can be defined as the number of association rules whose antecedents are tags in I_o and whose consequent is the candidate tag c. In other words, it is the number of "votes" a candidate tag has received from related tags associated with the target object.

5.1.2 *Descriptive power attributes*

Descriptive power attributes usually estimate the descriptive capacity of candidate tags based on statistics of their occurrence in the textual features of the target object. We [13] proposed the use of four of these attributes for tag recommendation. We start by defining the *Term Spread* of a candidate c in an object o, $TS(c, o)$, as the number of textual features (except tags if we desire to recommend only "new" tags for that object) of o that contain c [3].

The rationale behind $TS(c, o)$ is that the larger the number of textual features of o containing c, the more related c is to o's content. For example, if the term "X-men" appears in all features of a video, there is a high chance that the video is related to the famous comics. Our results in [3] indicate that, in isolation, TS provides better tag recommendations than the traditional TF in most datasets.

TF or *term frequency*, in turn, is the total number of occurrences of the candidate tag c in all textual features of the target object o and thus considers these textual features as a single bag of words. In contrast, TS takes into account the multiple textual blocks that compound the structure of the target object.

However, neither TS nor TF consider that some textual features may describe the content of the target object more accurately than others. For example, the title is usually the most representative textual feature of the object's content [3]. Thus, we proposed in [13] two other attributes, which extend TF and TS, weighting a candidate tag based on the average descriptive powers of the textual features in which it appears.

To define these new attributes, we need first to automatically estimate the descriptive power of a textual feature F_i using the *average feature spread* (AFS) metric [3]. Let the *feature instance spread* of a feature $F_{i,o}$ associated with an object o, $FIS(F_{i,o})$, be the average TS over all terms in $F_{i,o}$. We define $AFS(F_i)$ as the average $FIS(F_{i,o})$ over all instances of F_i associated with objects in the training set D. Thus, we define weighted TS (wTS) and weighted TF (wTF) as

$$wTS(c,o) = \sum_{F_{i,o} \in o} \mathbf{I}(c,F_{i,o}) \times AFS(F_i), \quad where\ \mathbf{I}(c,F_{i,o}) = \begin{cases} 1, if\ c \in F_{i,o} \\ 0, otherwise \end{cases}$$

$$wTF(c,o) = \sum_{F_{i,o} \in o} tf(c,F_{i,o}) \times AFS(F_i) \tag{2}$$

where $tf(c, F_{i,o})$ is the number of occurrences of the candidate tag c in textual feature $F_{i,o}$ of the target object o.

5.1.3 Discriminative power attributes

Discriminative power attributes promote more infrequent terms as tags, since they may better *discriminate* objects into different categories, topics, or levels of relevance, particularly considering that several services (e.g., classification, searching) often perform IR on multimedia content by using the associated tags as data sources. This aspect is captured by the *inverse feature frequency* (IFF) attribute [3], directly derived from the traditional *inverse document frequency* (IDF), considering, however, the term frequency in a specific textual feature (tags, in this case), instead of the full set of terms associated with the objects in the training dataset D. Given the number of elements in the training set $N = |D|$, the IFF of a candidate tag c in a textual feature i (tags in this case) is defined as $IFF(c, i) = log((N + 1)/(f_i(c) + 1))$, where $f_i(c)$ is the number objects in the training set in which c appears in the textual feature i. In our case, $f_i(c)$ is the number of training objects that are tagged with c.

We note that the value 1 is added to both numerator and denominator, without harming the tag specificity estimation, to deal with the value 0 in the denominator, which occurs for new terms that do not appear as tags in the training data.

IFF may have privilege terms from other textual features that do not appear as tags in the training data or noisy terms such as typos. Nevertheless, this attribute can be combined with the other attributes into a function, using, for example, learning-to-rank algorithms. Thus, its relative weight can be adjusted in order to avoid negative impacts in tag recommendation effectiveness.

Considering that both too general and too specific or noisy terms may not be ideal tag recommendations, [2] propose the *stability* attribute, which promotes terms with intermediate frequency values.

5.1.4 Term predictability

Another important aspect for tag recommendation is term predictability. Heymann et al. [22] measure this characteristic through the term's *entropy*.

If a term occurs consistently with certain tags, it is more predictable, thus having lower entropy. Terms that occur indiscriminately with many other tags are less predictable, thus having higher entropy. Term entropy can be useful particularly for breaking ties, as it is better to recommend more "consistent" or less "confusing" terms.

Another predictability attribute, called *Pred* [13, 18], measures the probability that a term is used as a tag in an object given that it was used in another textual feature of the same object.

5.1.5 User interest attribute

The user frequency (UF) attribute was used in [13, 18] in order to estimate the relevance of a candidate tag for a target user and thus provides personalized recommendations. *UF(c, u)* is simply the frequency at which the target user u assigns a candidate tag c to objects in a training collection. The idea is that the more frequently a user u assigns a candidate tag c to other objects in the application, the more relevant c is for u.

It is also common to exploit the temporal dynamics of tagging, particularly in user frequency-based tag attributes. From the observation that the temporal decay of the users' word choices follows a power-law function, the authors in [23] integrate a time component that gives more weight to tags that have been used more recently.

5.2 Learn-to-rank-based tag recommendation

Observing that recommendation is usually modeled as a ranking problem (i.e., we want to recommend the most relevant items first), learning-to-rank (L2R) techniques constitute an appropriate approach to tackle it. L2R-based methods are supervised approaches that automatically "learn" a ranking function from "previously seen" data known as training instances. Such training examples usually consist of candidate tags, their tag quality attribute values, and their relevance labels, which indicates their relevance levels. These labels can be assigned either manually or by exploiting previous tag assignments as ground truth. The objective of L2R approaches is to generate a model (function) that maps the tag quality attributes into a relevance score or rank.

More formally, for each candidate tag c for each object o (or pair object-user $<o, u>$ for personalized recommendation), we associate a vector $x_{c,o} \in \mathbb{R}^m$ (or $x_{c,o,u} \in \mathbb{R}^m$), where m is the number of considered tag quality attributes (e.g., each metric defined in Section 4). For training instances, we also assign a relevance label $y_{c,o}$ (or $y_{c,o,u}$), indicating the relevance level of candidate tag c to the object o (and user u). For example, we can define two relevance levels: 1 for relevant tags, and 0 for nonrelevant tags. In the offline training step, this data is exploited to generate the recommendation model. In the online recommendation step, in which we have new objects or users as input, the $y_{c,o}$ (or $y_{c,o,u}$) values are unknown, and the model learned in the training step is applied in order to predict these values.

Various L2R-based algorithms have been proposed for tag recommendation in the literature, including RankSVM, RankBoost, Genetic Programming, Random Forest (RF), Multiple Additive Regression Trees (MART), Lambda-MART, AdaRank, ListNet, Ranknet, and Coordinate Ascent. In [24] we can find a brief description of each of these algorithms and experimental results of the comparison of these methods using the RankLib tool (https://sourceforge.net/p/lemur/wiki/RankLib/). According to our results, RF, MART, and Lambda-MART are found to be the best performing strategies for the tag recommendation problem.

In [25], the author reviewed existing L2R algorithms in the context of document ranking, categorizing them into three approaches: pointwise, pairwise, and listwise. The *pointwise* approach associates a numerical score to each query-document pair and thus approximates the ranking problem by a regression problem. *Pairwise* approaches, in turn, transform the ranking problem into binary classification: given a pair of documents (or tags, in our case), we need to predict which one is the most relevant. Finally, the *listwise* approaches try to directly optimize a given evaluation measure.

Finally, it is also worth mentioning that, instead of adopting an attribute engineering approach, exploiting various handcrafted attributes like those described in Section 4, some recent works focus on investigating techniques that can learn attribute interactions from raw data, such as deep learning and factorization machines (FM) [26, 27]. The most representative method of this group is pairwise interaction tensor factorization (PITF). In this method, the tensor (i.e., a "tridimensional matrix") that models the pairwise interactions among users, items, and tags (i.e., the ranking preferences of the tags for each pair user object, which is obtained from the folksonomy relation data) is factored into lower-dimensional matrices to reduce noise [27]. The PITF model is learned from an adaption of the Bayesian personalized ranking (BPR) criterion. More recently, [26] exploit not only the folksonomy but also visual features of images, such as the objects appearing in the image, colors, shapes, or other visual aspects, into factorization machine models.

6. Conclusions

In this chapter, we have reviewed the main concepts related to tags and tag recommendation. There are various sources of data associated with Web 2.0 objects that can be used to extract and rank tags. Candidate tags can be extracted from the textual features associated with the target object using keyword extraction techniques, from mining co-occurrences with other tags, or other textual and content features, and from the neighborhood of the target object and/or target user. We also have briefly discussed various tag quality attributes that can be exploited to rank candidate tags. An effective way to combine these attributes is by means of learn-to-rank techniques, which can automatically "learn" tag recommendation functions from training examples.

Acknowledgements

Our research group is partially funded by Google, the Brazilian National Institute of Science and Technology for Web Research (MCT/CNPq/INCT Web Grant Number 573871/2008-6), and the authors' individual grants from CNPq, CAPES, and FAPEMIG.

Author details

Fabiano M. Belém*, Jussara M. Almeida and Marcos A. Gonçalves
Computer Science Department, Federal University of Minas Gerais, Belo Horizonte, Minas Gerais, Brazil

*Address all correspondence to: famube@gmail.com

References

[1] Gupta M, Li R, Yin Z, Han J. Survey on social tagging techniques. SIGKDD Explorations. **12**(1):58-72

[2] Sigurbjörnsson B, Zwol R. Flickr tag recommendation based on collective knowledge. In: Proceedings of WWW Conference. 2008. pp. 327-336

[3] Figueiredo F, Belém F, Pinto H, Almeida J, Gonçalves M. Assessing the quality of textual features in social media. Information Processing & Management. 2012;**49**:222-247

[4] Zuo Y, Zeng J, Gong M, Jiao L. Tag-aware recommender systems based on deep neural networks. Neurocomputing. 2016;**204**(C):51-60

[5] Hsu M, Chen H. Efficient and effective prediction of social tags to enhance web search. Journal of the Association for Information Science and Technology. 2011;**62**(8):1473-1487

[6] Oliveira V, Gomes G, Belém F, Brandão W, Almeida J, Ziviani N, et al. Automatic query expansion based on tag recommendation. In: Proceedings of the International Conference on Information and Knowledge Management. 2012. pp. 1985-1989

[7] Ribeiro I, Santos R, Gonçalves M, Laender H. On tag recommendation for expertise profiling: A case study in the scientific domain. In: Proceedings of the International Conference on Web Search and Data Mining. 2015. pp. 189-198

[8] Venetis P, Koutrika G, Garcia-Molina H. On the selection of tags for tag clouds. In: Proceedings of the ACM Conference on Web Search and Data Mining; 2011. pp. 835-834

[9] Belém F, Almeida J, Gonçalves M. A survey on tag recommendation methods. Journal of the Association for Information Science and Technology. 2016;**68**(4):830-844

[10] Spiteri L. Structure and form of folksonomy tags: The road to the public library catalogue. Webology. **4**(2):13-25

[11] Jäschke R, Marinho L, Hotho A, Schmidt-Thieme L, Stum G. Tag recommendations in folksonomies. In: Proceedings of the European Conference on Principles and Practice of Knowledge Discovery in Databases. 2007. pp. 506-514

[12] Wal V. Explaining and Showing Broad and Narrow Folksonomies. Retrieved on Oct 8th, 2015: http://www.vanderwal.net/random/entrysel.php?blog=1635

[13] Belém F, Martins E, Almeida J, Gonçalves M. Personalized and object-centered tag recommendation methods for web 2.0 applications. Information Processing & Management. 2014;**50**(4):524-553

[14] Belém F, Batista C, Santos R, Almeida J, Gonçalves M. Beyond relevance: Exploiting novelty and diversity in tag recommendation. ACM Transactions on Intelligent Systems and Technology. 2016;7:26:1-26:34

[15] Baeza-Yates R, Ribeiro-Neto B. Modern Information Retrieval. Boston, USA: Addison-Wesley; 2011

[16] Hulth A. Improved automatic keyword extraction given more linguistic knowledge. In: Conference on Empirical Methods in Natural Language Processing. 2003. pp. 216-223

[17] Rose S, Engel D, Cramer N, Cowley W. Automatic keyword extraction from individual documents. In: Text Mining: Applications and Theory. Wiley; 2010. pp. 1-20. Available at: https://onlinelibrary.wiley.com/doi/pdf/10.1002/9780470689646.ch1

[18] Lipczak M, Milios E. Efficient tag recommendation for real-life data.

ACM Transactions on Intelligent SystemsTechnology. 2011;**3**(1):2:1-2:21

[19] Wu L, Yang L, Yu N, Hua X. Learning to tag. In: Proceedings of the 18th International Conference on World WideWeb. 2009. pp. 361-370

[20] Zhu X, Nejdl W, Georgescu M. An adaptive teleportation random walk model for learning social tag relevance. In: Proceedings of the 37th International ACM SIGIR Conference on Research and Development in Information Retrieval. 2014. pp. 223-232

[21] Martins E, Belém F, Almeida J, Gonçalves M. On cold start for associative tag recommendation. Journal of the Association for Information Science and Technology. 2016;**67**(1):83-105

[22] Heymann P, Ramage D, Garcia-Molina H. Social tag prediction. In: Proceedings of the 31st International ACM SIGIR Conference on Research and Development in Information Retrieval. 2008. pp. 531-538

[23] Kowald D. Modeling cognitive processes in social tagging to improve tag recommendations. PHD-Symposium paper. 2018. Available at: https://arxiv.org/pdf/1805.11878.pdf

[24] Canuto S, Belém F, Almeida J, Gonçalves M. A comparative study of learning-to-rank techniques for tag recommendation. Journal of Information and Data Management. 2013;**4**:453-468

[25] Liu T-Y. Learning to rank for information retrieval. Foundations and Trends in Information Retrieval. 2009;**3**(3):225-331

[26] Nguyen H, Wistuba M, Schmidt-Thieme L. Personalized tag recommendation for images using deep transfer learning. In: Ceci M, Hollmen J, Todorovski L, Vens C, Dzeroski S, editors. Joint European Conference on Machine Learning and Knowledge Discovery in Databases. 2017. pp. 705-720

[27] Rendle S, Schmidt-Thieme L. Pairwise interaction tensor factorization for personalized tag recommendation. In: Proceedings of the Third ACM International Conference on Web Search and Data Mining. 2010. pp. 81-90

Chapter 10

Classification Model for Bullying Posts Detection

K. Nalini and L. Jabasheela

Abstract

Nowadays, many research tasks are concentrating on Social Media for Analyzing Sentiments and Opinions, Political Issues, Marketing Strategies and many more. Several text mining structures have been designed for different applications. Harassing is a category of claiming social turmoil in different structures and conduct toward a singular or group, to damage others. Investigation outcomes demonstrated that 7 young people out of 10 become the casualty of cyber bullying. Throughout the world, many prominent cases are existing due to the bad communications over the Web. So there could be suitable solutions for this problem and there is a need to eradicate the lacking in existing strategies in dealing problems with cyber bullying incidents. A prominent aim is to design a scheme to alert the people those who are using social networks and also to prevent them from bullying environments. Tweet corpus carries the messages in the text as well as it has ID, time, and so forth. The messages are imparted in informal form and furthermore, there is variety in the dialect. So, there is a requirement to operate a progression of filtration to handle the raw tweets before feature extraction and frequency extraction. The idea is to regard each tweet as a limited blend over a basic arrangement of topics, each of which is described by dissemination over words, and after that analyze tweets through such topic dispersions. Naturally, bullying topics might be related to higher probabilities for bullying words. An arrangement of training tweets with both bullying and non-bullying texts are required to take in a model that can derive topic distributions from tweets. Topic modeling is used to get lexical collocation designs in the irreverent content and create significant topics for a model.

Keywords: cyberbullying, Twitter, LDA, SVM, TF-IDF

1. Introduction

The proposed methodology is a dual compound method. It utilizes the arrangement of "bullying" or "non-bullying" class and also it utilizes link analysis to locate the most dynamic users as predators and victims. Each step can be explained in detail as follows. The feature selection is an essential phase in denoting data within component space to the categorizers. Mostly the data available from social network are noisy. So, there is a need to apply pre-processing techniques in order to obtain the research data with better quality followed by successive systematic steps; Moreover, sparsity in feature space increases with the count of documents. Nevertheless, the following types of features generated through the B-LDA topic model and weighted B-TF-IDF scheme. In the initial step, semantic highlights are related for locating harassing, abusive and offending posts. In pestering discovery the presence of pronouns in the nuisance post was represented. Essentially in this work, three sorts of capabilities are utilized. They are depicted as follows: (i) all second individual pronouns "you," "yourself," and so forth are considered one term; (ii) all other outstanding pronouns "he," "she," and so on., are viewed together as another element; (iii) foul words such as "fr**k," "shit," "moronic," and so forth., which make the post merciless are assembled in another arrangement of highlights. The new harassing words lexicon was made in view of the accompanying essential sites like *noswearing.com and urban dictionary*. The primary rationale behind consolidating these features is that it will boost the viability of the classification of tormenting posts. The classification outcomes are revealed in the experiments.

2. Review of literature

Rahat et al. [1] presented a multi-stage cyber bullying detection results that radically decreases the classification period and give warning signals. The system is greatly scalable without forfeiting precision and highly approachable in raising signals. It also contained an active priority scheduler and a rising classification procedure by applying Vine data sets. The performance outcomes demonstrate that the model enhances the scalability of digital harassing discovery contrasted to non-priority model and also explained that the system could fully check Vine-scale networks. The results depict that this digital harassing detection is considerably more measurable and receptive than the present modern technology. Zhong et al. [2] proposed an investigation to find out cyberbullying in Instagram utilizing the improvement of early-warning methods to detect offensive images.

The research operated by obtaining a huge volume of pictures in the Instagram image sharing process along with messages. They studied new features of the topics acquired from the picture portrayal and trained using neural network technology, added with images and texts. The results got the potential objectives for harassing on the characterization of texts and images. Sherly [3] proposed research using supervised feature selection to select the characteristics from the tweets by the ranking method. Then extreme learning machine (ELM) classifier is applied to execute the cyberbullying detection and enhance the precision and reduce the performance period. The performance investigation of the SFS-ELM model observed that the accuracy is improved by 13% and executed using MATLAB. Micheline et al. [4] accomplished a study by using an unsupervised methodology to identify harassing messages in social networks, utilizing Growing Hierarchical Self Organizing Map. The research contains various features to find semantic and syntactic interactions of regular cyber tormentors. They conducted various trials on FormSpring, Twitter and YouTube networks by collecting real time datasets. The outcomes of the research show that the model attains the significant performance and also promotes permanent watching applications to alleviate the huge issues of harassing. Suchini et al. [5] applied a text classification model to categorize the text as insulting or not. Feature selection is performed using Chi-square test and then classification algorithms are utilized for segregating comments as insulting or non-insulting words. Various algorithms like SVM, Naive Bayes, Logistic Regression, Random Forest are applied and out of all algorithms, SVM gave better results.

Krishna et al. [6] proposed a model deployed for detecting abusive text and images in the social network. This automated system could find the offensive content in messages using the combination of a bag of visual word method, local binary pattern and SVM classifier. The offensive detection in the text messages are executed by a bag of word method with Naïve Bayes classifier and then the Boolean system is applied to classify the content. Javier et al. [7] have displayed automatic strategies for identifying erotic plundering in Chat rooms. They have effectively demonstrated that a learning-based technique is an attainable method to approach this issue and have proposed novel sets of highlights to determine the classification of chat partakers as exploiters or non-exploiters. They exhibited that the arrangements of features used and the comparative weighting of the disarrangement expenditures in the SVMs are two fundamental factors that ought to be considered to upgrade execution.

Huang et al. [8] proposed normal text investigation using social network characteristics to classify harassing in Twitter and also considered the social connection between clients would betterment outcome for classification. Zhao et al. [9] applied a collection of features known as EBoW (Natural Language Processing method), containing a bag of words structure connected with Latent Semantic analysis and word embeddings by computing word vectors. They also used SVM to classify the data collection in Twitter which contains keywords like bully or bullying.

Chen et al. [10] researched existing content mining techniques in recognizing harassing texts for ensuring adolescent online safety. In particular, they proposed the Lexical Syntactical Feature (LSF) way to deal with hostile contents on the internet and further foresee a client's potentiality to convey hostile contents. Their investigation has many commitments. To begin with, they essentially conceptualize the idea of online hostile contents and further recognize the contribution of pejoratives/obscenities and profanities in deciding offensive substance, and present hand creating syntactic standards in finding verbally abusing provocation. Second, they enhanced customary Machine-Learning strategies by not just utilizing lexical features to identify hostile dialect, yet in addition style feature, structure features, and content-specific features to better foresee a client's possibility to convey hostile content in social media. Investigation result demonstrates that the LSF Sentence offensiveness forecast and client offensiveness estimate algorithm beat, customary learning-based methodologies in turns of precision, recall, and F-score. The LSF endures casual and incorrect spelling contents and it can possibly adjust to any forms of English written word styles.

3. The Bully-latent Dirichlet allocation (B-LDA): model design

LDA is an outstanding method of Bayesian multinomial mixture model in text analysis based on its ability to assemble, elucidate and semantically cogent topics. It uses the Dirichlet distribution to model the distribution of the topics for each and every one document. In LDA, each word is measured from a multinomial distribution over words particular to this topic. Since LDA is extremely modular and hierarchical, consequently, it can simply be broadened. Various expansions to basic LDA model have been recommended to incorporate document metadata. The easy process of integrating the metadata in generative topic models is to create both the words and the metadata concurrently specified unseen topic variables. The Author-Topic (AT) model resembles Bayesian network, in which every authors' attractions are modeled with a combination of topics [11]. In this model an arrangement of authors, advertisements are watched and looked over from different documents depends on their topics. To create each word, an author x is picked at identical from this set, then a topic z is chosen from a topic distribution θ_x that is particular to the author, and after that, a word w is created by testing from a topic-particular multinomial distribution ϕ_z.

The proposed Bully-LDA (B-LDA) model is used for identifying bullying words used by authors. This model captures bullying-topics which are used in social networks like Twitter. In Twitter, one person sends tweets to many followers. Here in this model, the sender is considered as Predator, when he/she sends bullying words to their followers. The followers are represented as Victims. The B-LDA model is a generative process model and also encapsulates topics and the communication networks of Predators and Victims by conditioning the multinomial distribution over bullying topics distinctly on both the Predator and a Victim of a bullying message. Unlike other models, B-LDA model takes into concern both predator and victims distinctly. The motive of the predator is also considered in addition to this representation. Each motive is associated with a set of topics, and these topics may overlap. For example, the categories of motive can be racist, sexual, outrage, irrelevant. The sexual motive of predator contains the topics of crude, implicit/ambiguous language or an indecent proposal. The Racist category contains more abusive matters such as homophobia, extremism, slurs, etc. The outrage is a category, which specifies reactions that express contempt. The messages that do not contain any form of offensive language are considered to be irrelevant. Each predator has a multinomial distribution over motives. Thus, B-LDA model is a clustering model, in which appearances of topics are the underlying data, and sets of correlated topics are together gathered as clusters that denote motive. Predators and Victims are mapped to motive assignments, and then a topic is selected based on these motives. The intention of each and every predator has a multinomial distribution on topics, and every topic has a multinomial distribution on words. First, the motive assignments can be made separately for each word in a document. This model represents that someone can change motive during the exchange of the messages.

Author-Topic (AT) [11] model has been extended by incorporating a new set of variables like authors as Predators and Victims, the motivation of an author. In this generative process for each message, a Predator, p_d and a set of Victims, v_d are observed. To generate each word, a victim y is chosen at uniform from v_d, and then a motive x for the Predator is chosen from multinomial motive distribution ψpd. Next a topic z is selected from a multinomial topic distribution θ_x, in which the

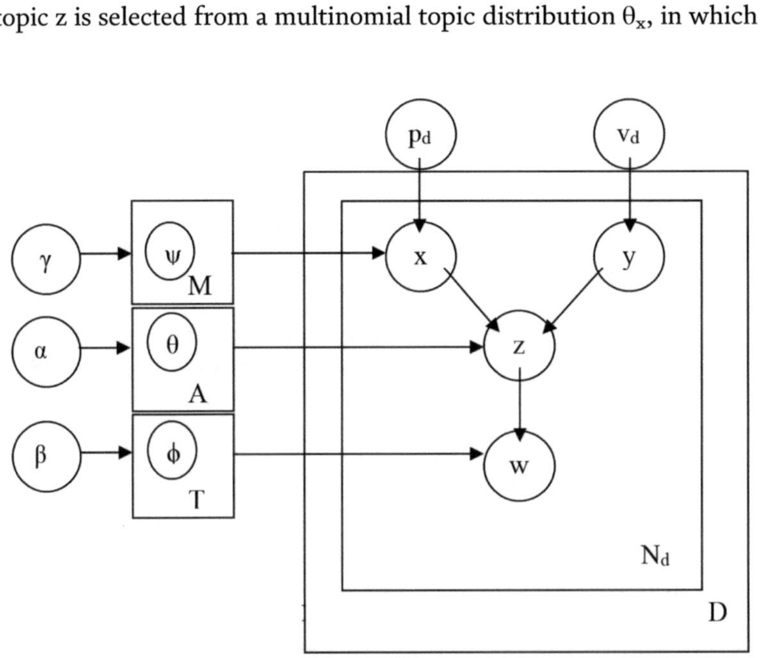

Figure 1.
Graphical model for B-LDA.

distribution is specific to the predator-motive(x). At last, the word w is produced by sampling from a topic-meticulous multinomial distribution ϕ_z.

Figure 1 is a schematic diagram of the B-LDA model.

The generative procedure of this strategy is as follows:

1. for every motive m with m = 1, … ..M, choose $\psi_m \sim Dir(\gamma)$
2. for each predator and victim pair (x,y) with x = 1, … .,A and y = 1, … .,A choose $\theta x,y \sim$ Dir (α)
3. for each topic t with t = 1, … ..T, choose $\phi_t \sim Dir(\beta)$
4. for each message d
    a. observe motive m_d
    b. observe predator p_d and the victims v_d
    c. for each word w in d
        i. choose topic $z_{dn} \sim \theta_{zd}$
        ii. choose word $w_{dn} \sim \phi_{zdn}$

In this model for a particular message d, given the hyper parameters α, β, and γ, the predator p_d, and set of victims v_d, the connected dispersion of an author blend θ, a motive blend ψ, a topic blend ϕ, a set of N_d victims y_d, and a set of N_d predator motives x_d, a set of N_d topics z_d and a set of N_d words w_d is assigned by,

$$\begin{aligned} p(\theta,\phi,\psi,yd,xd,zd,wd|\alpha,\beta,\gamma,pd,vd) \\ = p(\psi|\gamma)p(\theta|\alpha)p(\phi|\beta) \\ = \prod_{n=1}^{Nd} p(ydn|vd)p(xdn|pd)p(zdn|\theta xdn)p(wdn|\phi zdn) \end{aligned} \tag{1}$$

Integrating over γ, θ and ϕ and summing over y_d, x_d, and z_d, the marginal distribution of a document is calculated as follows:

$$\begin{aligned} p(wd|\alpha,\beta,\gamma,pd,vd) = \iiint p(\psi|\gamma)p(\theta|\alpha)p(\phi|\beta) \\ \prod_{n=1}^{Nd}\sum_{ydn}\sum_{xdn}\sum_{zdn} p(ydn|vd)p(xdn|pd)p(zdn|\theta xdn)p(wdn|\phi zdn)d\psi d\phi d\theta \end{aligned} \tag{2}$$

Then the product of the marginal probabilities of single documents, and the probability of a corpus is computed as,

$$p(D|\alpha,\beta,\gamma,p,v) = \prod_{d=1}^{D} p(wd|\alpha,\beta,\gamma,pd,vd) \tag{3}$$

3.1 Monte Carlo Gibbs sampling

The assumption on models in the LDA family cannot be carried out correctly. Three standard approximations have been occupied to acquire practical results:

Variational methods [12], Gibbs sampling [13], and expectation propagation [14]. As Gibbs sampling is easy to implement, it has been applied here. There is a need to derive a formula to carry out the Gibbs sampling for $P(z_i,y_i,x_i|z_{-i},y_{-i},x_{-i})$, the conditional distribution of a topic and victims for w word given all other words topic and victim assignment, the motive of the predator, z-i, y-i, and x-i. In order to calculate P(z,y,x|w), the posterior distribution of topic, victim assignments and the motive of the predator given the words in the corpus.

The calculations begin with P(w|z,x), using P(w|z,x,Φ) in order to integrate out the unknown Φ distributions to obtain: $P(w|z,y,\Phi)=\prod_{iw=1}^{W}\phi ziw(Wiw)$.

Reorganizing the product over the W word token exist in the corpus to collect words that are assigned to the same bullying topic,

$$P(w|z,y,\Phi)=\prod_{z=1}^{T}\prod_{u=1}^{U}\phi_z^{n_z{}^{wu}} \tag{4}$$

where $n_z{}^{wu}$ is the number of times that a bullying word, w_u was assigned to a bullying topic. To integrate out the ϕ distribution by using the Dirichlet distributions,

$$\begin{aligned} p(w|z,y) &= \int\prod_{z=1}^{T}\left(\frac{\Gamma\left(\sum_{u=1}^{U}\beta u\right)}{\prod_{u=1}^{U}\Gamma(\beta u)}\left(\prod_{u=1}^{U}\phi_z^{n_z^{wu}+\beta u-1}(wu)d\phi_z(wu)\right)\right) \\ &= \prod_{z=1}^{T}\left(\frac{\Gamma\left(\sum_{u=1}^{U}\beta u\right)}{\prod_{u=1}^{U}\Gamma(\beta u)}\left(\frac{\prod_{u=1}^{U}\Gamma\left(n_z^{wu}+\beta u\right)}{\Gamma\left(\sum_{u=1}^{U}\beta u+\sum_{u=1}^{U}n_z^{wu}\right)}\right)\right) \end{aligned} \tag{5}$$

In the same manner, P(z,y) is computed using a procedure analogous to that used for P(w|z,y). The collected terms of bullying words are assigned to the same topic and predator-victim pair and integrate out the Θ distributions corresponding to all the different predator-victim pairs, P:

$$P(z,y)=\left(\prod_{iw=1}^{W}\frac{1}{n_R(diw)}\right)\prod_{p=1}^{P}\left(\frac{\Gamma\left(\sum_z\alpha z\right)}{\prod_{z=1}^{T}\Gamma(\alpha z)}\frac{\prod_z\Gamma\left(n_p^z+\alpha_z\right)}{\Gamma\left(\sum_z\alpha_z+\sum_z n_p^z\right)}\right) \tag{6}$$

where $nR(diw)$ is the number of victims corresponding to a word in a message.

Similarly can calculate P(z, x) using a procedure analogous to that used for P (w|z, x). Bullying words have been assigned to the same topic and the motivation of the predator can be computed as,

$$P(z,x)=\left(\prod_{iw=1}^{W}\frac{1}{n_S(diw)}\right)\prod_{p=1}^{P}\left(\frac{\Gamma\left(\sum_z\gamma_z\right)}{\prod_{z=1}^{T}\Gamma(\gamma z)}\frac{\prod_z\Gamma\left(n_m^z+\gamma_z\right)}{\Gamma\left(\sum_z\gamma_z+\sum_z n_m^z\right)}\right) \tag{7}$$

where $nS(diw)$ is the number of predators having bad motivation with respect to the bullying word in a message. An expression for P (w, z, y, x) can be achieved by combining the equations of P(w|z, y), P(z, y) and P(z, x). This can be used to write an expression for the posterior distribution of z, y and x given the corpus,

$$P(z,y,x|w)=\frac{P(w,z,y,x)}{\sum_{z,y,x}P(w,z,y,x)} \tag{8}$$

Hence the denominator cannot be calculated directly. The following equations are used to run a MCMC Gibbs sampling calculation by using the conditional distribution $P(z_i, y_i, x_i, w_i|z_{-i}, y_{-i}, x_{-i}, w_{-i})$.

$$
\begin{aligned}
&P(zi,yi,xi,wi|z-i,y-i,x-i,w-i)\\
&=\frac{P(z,y,x,w)}{P(z-i,y-i,x-i,w-i)}\\
&=\frac{1}{nR}\left(\frac{\frac{\Gamma(n_m^t+\gamma t)}{\Gamma(\sum_z n_m^z+\sum_z \gamma z)}\frac{\Gamma\left(n_p^t+\alpha t\right)}{\Gamma\left(\sum_z n_p^z+\sum_z \alpha z\right)}\frac{\Gamma(n_t^{wu}+\beta u)}{\Gamma(\sum_u n_t^{wu}+\sum_u \beta u)}}{\frac{\Gamma(n_m^t-1+\gamma t)}{\Gamma(\sum_z n_m^z-1+\sum_z \gamma z)}\frac{\Gamma\left(n_p^t-1+\alpha t\right)}{\Gamma\left(\sum_z n_p^z-1+\sum_z \alpha z\right)}\frac{\Gamma(n_t^{wu}-1+\beta u)}{\Gamma(\sum_u n_t^{wu}-1+\sum_u \beta u)}}\right)\\
&=\frac{1}{nR}\frac{n_{m,-i}^t+\gamma t}{\sum_z n_{m,-i}^z+\sum_z \gamma z}\frac{n_{p,-i}^t+\alpha t}{\sum_z n_{p,-i}^z+\sum_z \alpha z}\frac{n_{t,-i}^{wu}+\beta u}{\sum_u n_{t,-i}+\sum_u \beta u}
\end{aligned}
\tag{9}
$$

where the victim, y is part of Predator-Victim pair, p, the –i subscript is used to denote that the counts are taken by excluding the assignment of word i itself, and n_R is the number of Victims for the message to which word i belongs.

3.2 Experiments and results

In this chapter, the experimental results are discussed. The datasets used in these experiments are tweets from Twitter. An experiment has been conducted on tweets based on the architecture of an automatic cyber bullying detection system. Search is made in the Twitter stream for Tweets containing the strings that contain offensive words so as to particularly filter for tweets related to bullying. In total, more than 1,00,000 tweets are gathered between Jan 1st, 2015 and Jan 30th, 2016. A limit number of tweets are matching with the query. So, approximately 300 tweets are filtered per day. The statistics for training and the testing corpus is given in **Table 1**. Tweets were manually labeled as belonging to one of the different motives namely Sexual, Racist, Outrage, Irrelevant, and Unknown after the preprocessing. The examples of harassing comments posted on Twitter are listed below and depicted in **Figure 2(a)** and **(b)** and top bullying words which are extracted are given in **Table 2** (**Figures 3–5**).

Date	Time	Tweets
01–13-15	12:16	NefarioussNess Do not fuck with people's hearts
09–18-15	11:51	TittyCityClay it's always been a self respect thing. Shit like this is stupid as fuck lol
05–13-15	10:11	djkeneechi Nah kiss no one ass to stay in my life anymore im tired of that shit it's time for me to man up

3.3 Results and discussions

Bully-Latent Dirichlet Allocation model is an intended for pictorial representation of texts in a harassing message, given their predator and a pair of casualties.

	Training corpus	Testing corpus
Tweets	3,18,14,716	97,35,537
Retweets	76,20,335	2,87,567
URLs	85,45,112	4,76,234
Usernames	97,02,445	14,20,554
Hashtags	79,85,956	3,56,778

Table 1.
Statistics of training and testing corpus.

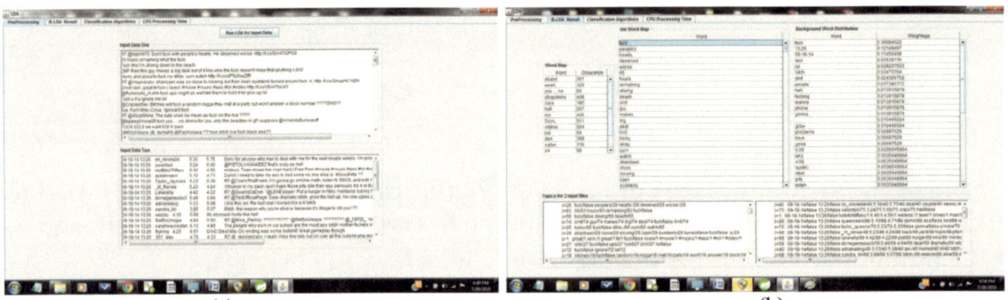

(a) (b)

Figure 2.
(a) Bullying words with their probability, and (b) List of bullying words.

Word	Prob	Word	Prob	Word	Prob
Fuck	0.0798	Bitch	0.0705	Naked	0.0588
Ass	0.0767	Freak	0.0699	Sexy	0.0569
shit	0.0752	Fat	0.0663	Mood	0.0547
Gay	0.0738	Dirty	0.0643	Lick	0.0519
Dumb	0.0722	Bullshit	0.0621	Bed	0.0508
Suck	0.0711	Kiss	0.0604	Piss	0.0495

Table 2.
Extracted top bullying words.

Figure 3.
Word cloud for bullying words.

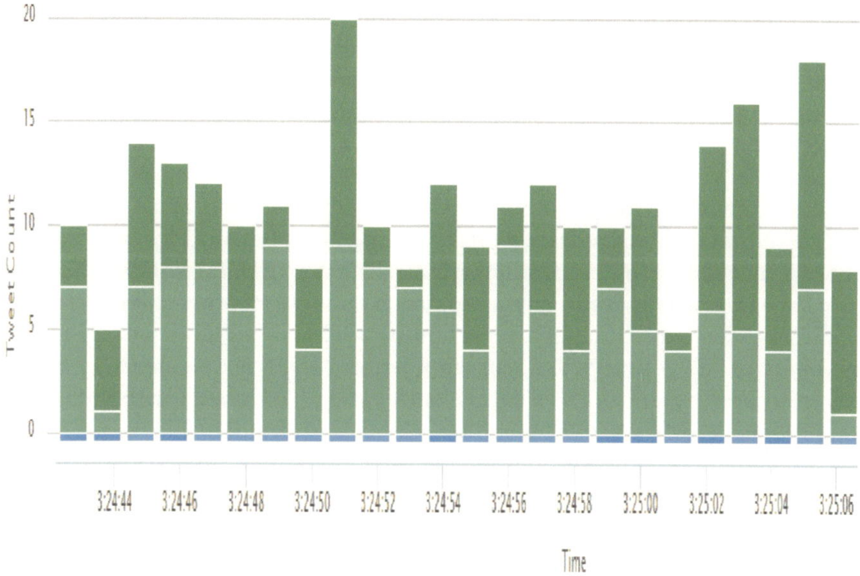

Figure 4.
Number of bullying tweets over time intervals.

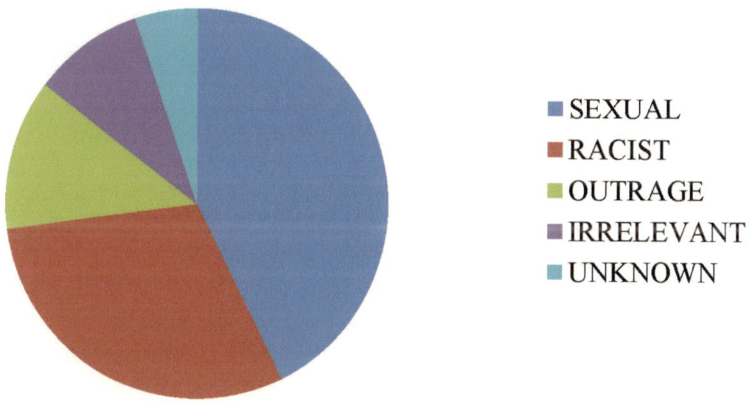

Figure 5.
Distributions of tweets per motive.

B-LDA got crucial enrichment to facilitate specification the per-bullying message topic dispersion mutually on the predator and individual victims. Every topic includes multinomial distribution on words and every Predator-Casualty pair has a distribution on topics. So, subsidiary dispersions in excess on bullying subjects accustomed exclusively on a predator, or solely on a recipient, can be computed easily. For example, corpus comprising 135 persons and 35 k bulling messages, and also on 5 months of sending and receiving messages of a predator, comprising 17 victims and 19 k messages. B-LDA turns up tremendously prominent topics, and grants support that it predicts predator's motives. In the experiments, the hectic parameters α and β are fixed at 1 and 0.01 respectively. The number of topics T is also fixed at $|T| = 5$. For a 50 topic solution, Dataset from Twitter took 150 hours for 2000 iterations (5 min per iteration).

B-LDA proves the motive of the predator and track the activity of the predator with victims, using the following steps. First, the proportions of each predator contributing in each of the bullying topics are determined. Next the impacts of the predators throughout the time intervals on the bullying topics. The two users' threshold ε and λ are empirically set to 3.2106 and 2.0457, respectively. From each of the documents, B-LDA generates 5 topics with predators associated with each. The distribution of the different bullying topics from the documents is displayed in **Table 3**. From the table, predator p1 has a probability of 0.0547 for bullying topic t5. There is a need to prove the bullying motive of the predator with victim using specific time intervals within bullying topics. It could be characterized as trails: A tweet message is a triplet (a, μ, τ), representing a textual bullying message μ written by the predator "a" at time τ. A document, denoted by d, is a sequence of bullying messages ordered by τ. From this definition, time τ_d is associated with both message μ_d and predator a_d.

The predator time contributions during time interval have been evaluated by:

$$F\left(a_d^t\right)_{\mathrm{T}^s}^{\mathrm{T}^f} = \begin{cases} active & if\ p\left(a_d^t\right)_{\mathrm{T}^s}^{\mathrm{T}^f} \geq users\ threshold, F(t)_{\mathrm{T}^s}^{\mathrm{T}^f}\ is\ active \\ not\text{-}active & otherwise \end{cases} \tag{10}$$

A predator is said to be active and his/her motive of bullying during the interval $[\tau^s, \tau^f]$ for topic t if the probability of a predator participating in t, during that time period, exceeds the user-specified threshold, and $F(t)_{\mathrm{T}^s}^{\mathrm{T}^f}$ is active within that duration. The user enumerated threshold is calculated by taking an average of ϑ_a^t over predators for t. The contribution of a predator $a_{i,d}^t$ within $[\tau^s, \tau^f]$, using $P\left(a^{\mathrm{T}^s}|t\right) = \frac{p\left(a^{\mathrm{T}^s}|d^{\mathrm{T}^s}\right) \times p\left(t^{\mathrm{T}^s}|d^{\mathrm{T}^s}\right)}{p\left(d^{\mathrm{T}^s}\right)}$ per tome instance s, is mapped first in order to compute $p\left(a_{i,d}^t\right)_{\mathrm{T}^s}^{\mathrm{T}^f}$. Next, the total probability for predator a^t during $[\tau^s, \tau^f]$ is calculated as $\sum_{\mathrm{T}^s}^{\mathrm{T}^f} P\left(a^{\mathrm{T}^s}|t\right)$. **Figure 6** shows the activity of predators over time. For example, the activity of predators in bullying topic $t_{5,d5}$ during [15:00,21:00] can be analyzed in the following manner. Initially, the specified threshold is determined as 0.1770, for the average of ϑ_a^t. Then the mapping function is calculated for all predators. For example, a predator a_5 and time instance s = 15:00 are considered to analyze. The mapping function is calculated as $P(a_{5,\mathrm{T}}15:00|t_5) = 0.0547$ and then the total probability of a_5 is estimated by calculating $\sum_{\mathrm{T}15:00}^{\mathrm{T}21:00} P(a_{5,\mathrm{T}^s}|t_5) = 0.2307$. When applying the transition function $F\left(a_d^t\right)_{\mathrm{T}^s}^{\mathrm{T}^f}$, the predators (a_1,a_3) are active for bullying topic $t_{5,d1}$ and the predators (a_2,a_4,a_5) are not active.

3.4 Performance evaluation

The Perplexity of the model is used on test documents to estimate the execution of model and it is a customary measure for evaluating the operation of a probabilistic model. The adapted models are compared by means of perplexity on test datasets. Perplexity is extensively used in a probabilistic model for checking their quality. The perplexity of a couple of trial texts, (w_d,p_d) for $d \in D^{test}$, is characterized as the exponential of the negative standardized predictive likelihood underneath the representation,

$$perplexity(wd|pd) = \exp\left[-\frac{\ln p(wd|pd)}{Nd}\right] \tag{11}$$

MOTIVE = RACISM									
TOPIC 5		TOPIC 10		TOPIC 15		TOPIC 20		TOPIC 25	
EXTREMISM		HOMOPHOBIA		VIOLENCE		REF. TO HANDICAPS		SLURS	
Incorrect	0.0271	ColdSweat	0.0265	Shit	0.0752	Fuck	0.0798	Pussi	0.0321
Improper	0.0242	Dread	0.0254	Bullshit	0.0621	Ass	0.0767	Dog	0.0312
Indecent	0.0231	Fearful	0.0235	Piss	0.0506	Dumb	0.0722	Filthy	0.0304
Ineligible	0.0225	Horror	0.0223	Aggrieve	0.0254	Blind	0.0342	Crow	0.0294
Unfit	0.0214	Panic	0.0212	Tee toe	0.0232	Cracy	0.0212	Nitchie	0.0276
Unsuited	0.021	Phobia	0.0203	Nose	0.0215	Daft	0.0203	Peckerwood	0.0253
Room	0.0197	Scare	0.0194	Gotoofar	0.0201	Autism	0.0167	Cameljockey	0.0238
Raffish	0.0193	Terror	0.0187	Rufflesb's feathers	0.0176	Freak	0.0154	Nigger	0.0221
Square peg	0.0184	Alarm	0.0176	Aggravate	0.0154	Gimpy	0.0132	Peckerwood	0.0213
Unworthy	0.0173	Fright	0.0169	Burn	0.0132	Windowlicker	0.0121	Wigger	0.0201
Predators: Victims	Prob	Predators: Victims	Prob	Predators: Victims	Prob	Predators: Victims	Prob	Predators: Victims	Prob
P1: V1	0.0547	P1: V1	0.0341	P4: V4	0.0352	P3: V5	0.0421	P1: V6	0.0284
P2: V2	0.0367	P2: V2	0.0288	P1: V2	0.0254	P2: V6	0.0325	P5: V7	0.0257
P3: V3	0.0361	P3: V3	0.0254	P1: V3	0.0246	P1: V3	0.0208	P4: V5	0.0236
MOTIVE = SEXUAL									
TOPIC 30		TOPIC 35		TOPIC 40		TOPIC 45		TOPIC 50	
CRUDE LANGUAGE		IMPLICIT LANGUAGE		INDECENT PROPOSALS		UNREFINED LANGUAGE		SLANG WORDS	
Gay	0.0738	Dirty	0.0643	Mood	0.0547	Bitch	0.0705	Pull	0.0456
Suck	0.0711	Bed	0.0508	Lick	0.0519	Freak	0.0699	Bumpuglies	0.0423
Naked	0.0588	Frequent	0.0491	Kiss	0.0508	Fat	0.0663	Fug	0.0321

MOTIVE = SEXUAL									
TOPIC 30		**TOPIC 35**		**TOPIC 40**		**TOPIC 45**		**TOPIC 50**	
CRUDE LANGUAGE		**IMPLICIT LANGUAGE**		**INDECENT PROPOSALS**		**UNREFINED LANGUAGE**		**SLANG WORDS**	
Sexy	0.0569	Sleep	0.0282	Hangnow	0.0485	Happyhappy	0.0341	Randy	0.0307
Kickit	0.0445	Kneedeep	0.0241	Givebusiness	0.0465	Poundduck	0.0324	Juicy	0.0284
FuckforOL’	0.0432	Encounter	0.0215	Monkeylove	0.0328	Homerun	0.0307	Hempedup	0.0245
Getdown dirty	0.0421	Donasty	0.0208	Sexytime	0.0319	Smack	0.0284	Jiffystiffy	0.0209
Slap	0.0316	doublebag	0.0165	Intimacy	0.0206	Serve	0.0271	Ride	0.0154
Hump	0.0307	Giveitup	0.0154	Cottage	0.0191	Jellosex	0.0135	Smush	0.0124
Screw	0.0201	Getlucky	0.0142	Raunchy	0.0147	Score	0.0104	Trim	0.0107
Predators: Victims	Prob	Predators: Victims	Prob	Predators: Victims	Prob	Predators: Victims	Prob	Predators: Victims	Prob
P1: V1	0.0737	P4: V4	0.0541	P3: V5	0.0452	P1: V 6	0.0595	P3: V5	0.0354
P2: v2	0.0552	P1: V2	0.0428	P2: V6	0.0321	P5: V7	0.0467	P2: V6	0.0241
P3: V3	0.0324	P1: V3	0.0367	P1: V3	0.0276	P4: V5	0.0354	P1: V3	0.0211

MOTIVE = OUTRAGE		MOTIVE = IRRELEVANT		MOTIVE = UNKNOWN	
TOPIC 60		**TOPIC 70**		**TOPIC 90**	
ANGER		Make out	0.0267	Outhouse	0.0246
Bitterness	0.0365	Marquee	0.0235	Pant	0.0232
Hard	0.0354	Mate	0.0223	Pass out	0.0214
Storm	0.0321	Minor	0.0215	Patient	0.0208
Irritation	0.0306	Moot	0.0209	PC	0.0179
Wrath	0.0268	MP	0.0201	Period	0.0165
Fury	0.0251	MUM	0.0189	Plant	0.0152

DOI: http://dx.doi.org/10.5772/intechopen.88633

MOTIVE = OUTRAGE		MOTIVE = IRRELEVANT		MOTIVE = UNKNOWN	
TOPIC 60		TOPIC 70		TOPIC 90	
ANGER					
Resent	0.0237	Nappy	0.0154	POP	0.0143
Rancor	0.0209	Natter	0.0142	Restroom	0.0137
Grudge	0.0192	Nick	0.0126	Rider	0.0129
Flap	0.0163	Nonce	0.0118	Sick	0.0109
Predators: Victims	Prob	Predators: Victims	Prob	Predators: Victims	Prob
P1: V7	0.0241	P5: V2	0.0207	P2: V6	0.0175
P2: V4	0.0219	P4: V5	0.0165	P5: V8	0.0154
P3: V3	0.0147	P1: V3	0.0126	P4: V5	0.0132

Table 3.
The distribution for the different bullying topics from the documents.

Better simplification functioning is designated by means of a lesser perplexity on a held-out document. The derivation of the likelihood of a collection of texts specified the predator is a uncomplicated computation in Bully-LDA model.

$$p(wd|pd) = \int d\theta \int d\phi p(\theta|Dtrain)p(\phi|Dtrain) * \prod_{m=1}^{Nd} \left[\frac{1}{Ad} \sum\nolimits_{i \in pd,j} \theta ij \phi wmj \right] \quad (12)$$

The term in the brackets is merely the probability for the word w_m specified the pair of predators p_d. The detailed results are exposed in **Figure 7**. These results

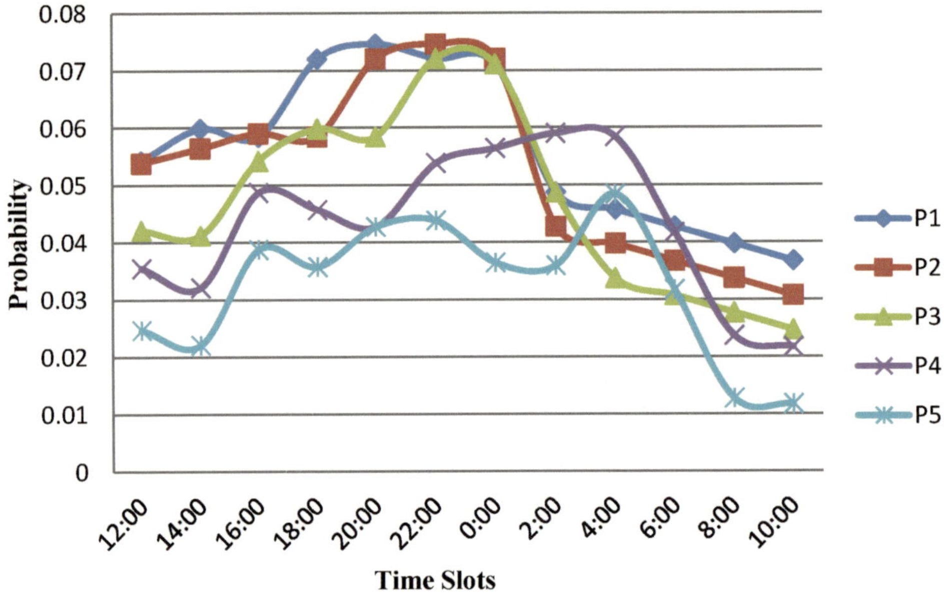

Figure 6.
Predators activity for bullying topic 30.

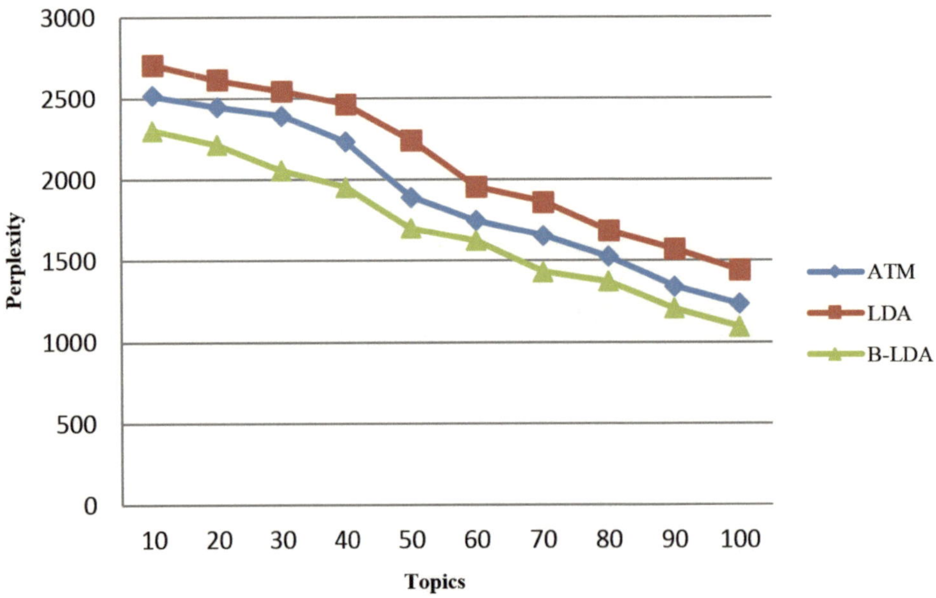

Figure 7.
Comparisons of different models in terms of perplexity.

indicate that B-LDA better generalizes performance than ATM and LDA. The improvement in generalization performance of B-LDA can be explained by its ability to better model when comparing with LDA and ATM model. If a word which has small probability in the bullying topics of training document, then it will cause an increase in perplexity. As the number of bullying topics increase, then the probabilities assigned to words get smaller in each bullying topic. Even though ATM models the roles of authors, does not show promising results and it is originally designed for the scenario where each document has multiple authors. It is clear that B-LDA achieves superior performances among all the adopted models. The perplexity of LDA, ATM, and B-LDA are closer and they decrease steadily with the increase of topics. According to human judgments, perplexity is not easy to correlate the results. So, it is necessary to compare the models using simple metrics like Precision, Recall, and F1 measure. The standard supervised classifier, i.e., Support Vector Machine (SVM), is adopted with B-LDA for classification. LibSVM was applied to the two-class classification problem using a linear kernel. Each post is an instance; positive classes contain bullying messages and negative classes contain non-bullying messages. A 10-fold cross-validation was performed in which the complete dataset was partitioned 10 times into 10 samples; in every round, nine portions were employed for exercising and the enduring section was applied for trial (**Figure 8**).

The functioning of the classifier was appraised on precision, recall and F-1 measure and these measures depend on the top-ranked features produced through B-LDA method against the truth set as tested on the datasets. Precision: The Aggregate number of accurately distinguished genuine harassing posts out of recovered tormenting cases. Recall: Number of effectively distinguished tormenting cases from an aggregate number of genuine harassing cases. F-1 measure: the equally weighted harmonic mean of precision and recall. **Table 4** shows the classifier performance.

3.5 Comparison of weighted B-TFIDF with baseline method

The weighted B-TFIDF method is compared with the work done in a content analysis in a web on four different datasets. The new feature selection method using weighted B-TFIDF proved that it is better than baseline. The outcomes are cataloged in **Table 5** and also indicate a very high precision, recall and F-1 measure on Twitter. In Kongregate precision fell down at the top 2000 features. In most of

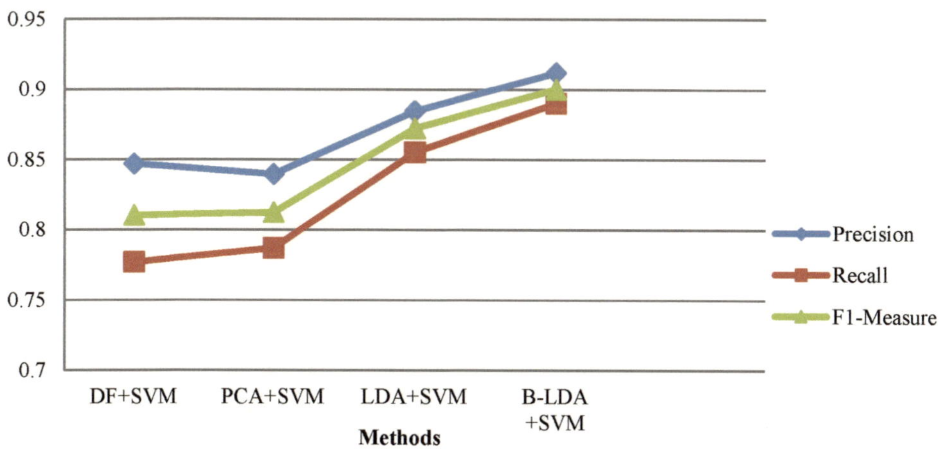

Figure 8.
Classifier performances based on different feature reduction methods.

the cases, the classifier performed almost similar, that is between 80 and 100%. On Myspace dataset recall is moderate nearing to 1. However, precision varies between 76 and 87% except at feature value 18,000 when it reaches 91%. Unlike other datasets, Slashdot performance is very low. Although recall is moderate, precision and F-1 measures decomposed while component set was low. Also, poor performance is observed at feature value 18,000. From this discussion, the performance of weighted B-TFIDF shows the best result (**Figure 9**).

3.6 Victim and predator identification

In order to identify cyber bullying predators and victims, there is need to determine the most active predators and the most attacked users. The most dynamic predators and victims, and look at the association of clients in a tormenting relationship as appeared in **Table 6** and it demonstrates that now and again there is more than one user at a similar rank. In this manner, users with a similar rank are gathered together. So it is important to notice that predators hailed at Rank I are additionally recognized as a victim at Rank II. Additionally, Rank II predators are Rank VII victims as well (**Figure 10**).

3.6.1 Graph representation

The major goal of a users' communication network are considered to identify predators and casualties. Gephi [15], a graphical interface is employed to monitor a user's link in the harassing posts in a network. **Figure 11** delineates the bullying network and it represents that a group of users obtained depend upon on the

Method	Precision	Recall	F-Measure
DF + SVM	0.8471	0.7770	0.8105
PCA + SVM	0.8397	0.7870	0.8125
LDA + SVM	0.8846	0.8554	0.8724
B-LDA + SVM	0.9121	0.8901	0.9003

Table 4.
Classifier performances based on different feature reduction methods.

		Kongregate	Slashdot	MySpace	Twitter
Baseline	Precision	0.35	0.32	0.42	0.62
Baseline	Recall	0.60	0.28	0.25	0.53
Baseline	F-1 measure	0.44	0.30	0.31	0.57
Weighted TFIDF	Precision	0.87	0.78	0.86	0.87
Weighted TFIDF	Recall	0.97	0.99	0.98	0.75
Weighted TFIDF	F-1 measure	0.92	0.87	0.92	0.81
Weighted B-TFIDF	Precision	0.95	0.96	0.96	0.98
Weighted B-TFIDF	Recall	0.93	0.84	0.93	0.96
Weighted B-TFIDF	F-1 measure	0.94	0.90	0.95	0.97

Table 5.
Comparison of weighted B-TFIDF with baseline method on other datasets.

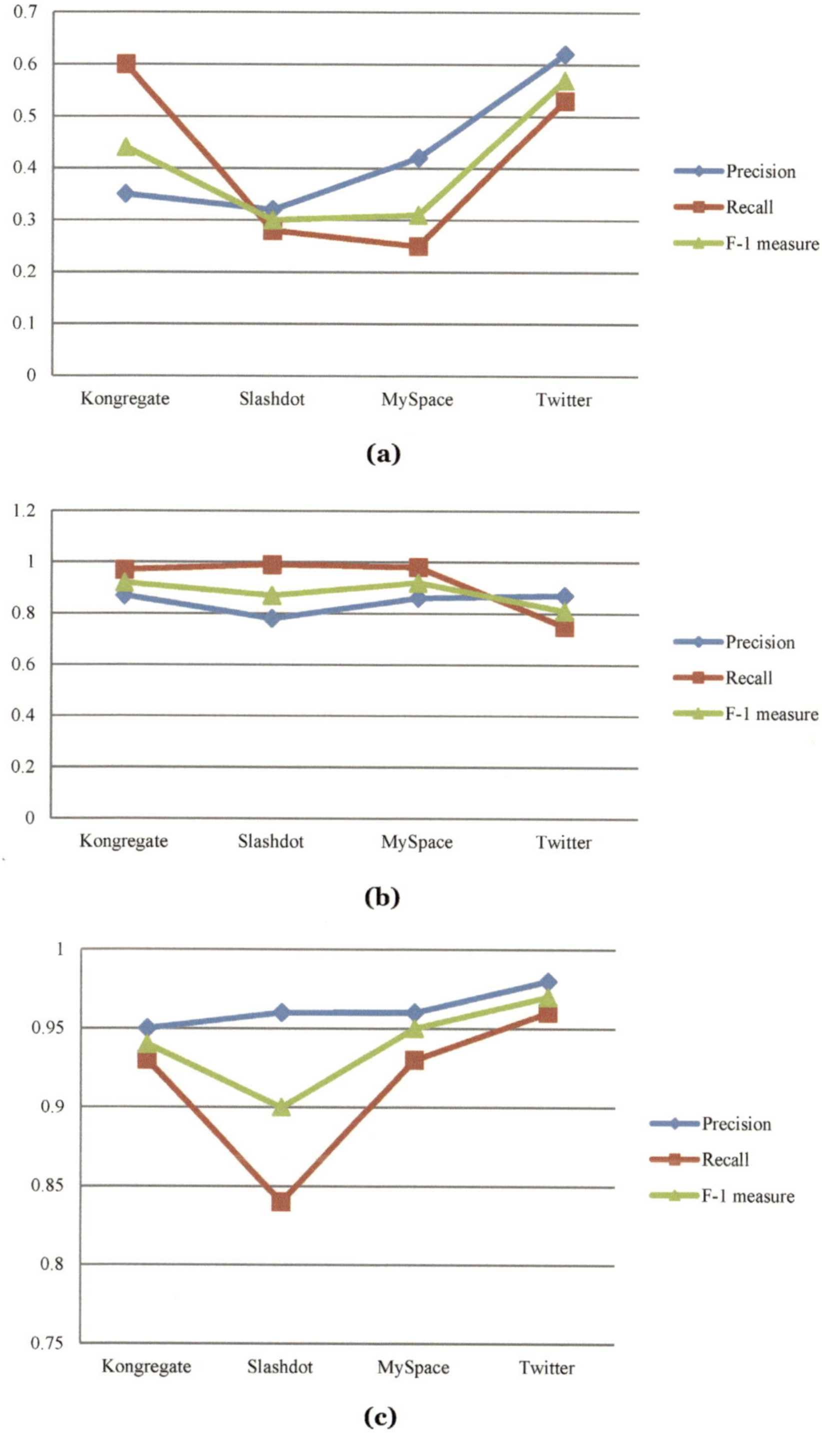

Figure 9.
(a) Base line method, (b) weighted TFIDF method, and (c) weighted B-TFIDF method.

tormenting messages by utilizing modularity theorem, in order to quantify the quality of segment of a system into sub-graphs or groups. Modularity is characterized as the summation of the weight of all the edges that sink inside the given subgroups less the expected part if edges were dispensed at arbitrary in a given graph.

Rank	I	II	III	IV	V	VI	VII	VIII
Number of users (predators)	4	2	1	1	2	7	3	2
Number of users (victims)	8	4	7	2	2	1	9	8

Table 6.
Performance of graph model: Predators and victims identification.

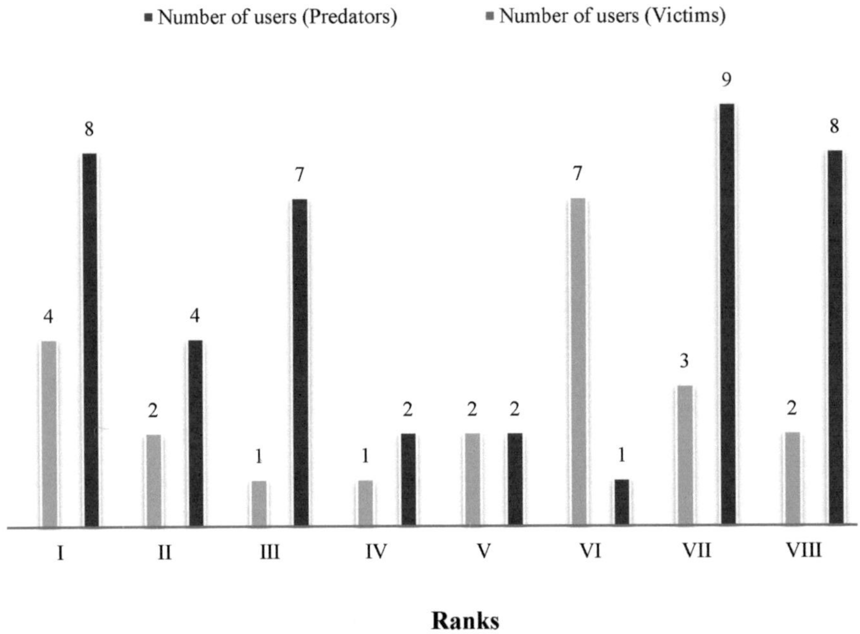

Figure 10.
Predators and victims identification.

As appeared in **Figure 11**, nine groups or communities, delineated by various colors are formed by considering users that are thickly connected inside the group contrasted with between group by utilizing modularity algorithm. The density of post indicates the badness embedded inside the post and it is calculated for each post. The thickness of a post is computed as the aggregate count of the harassing words within the post separated by the aggregate number of the words in the post. The HITS algorithm is utilized in order to recognize the predators and related casualties and it is also helpful to calculate their scores. The objective behind the HITS strategies is that in a network, the good hub pages point to good authorized pages which are connected by the good hub pages. The search query enters through web pages to recognize potential hub and authority pages with respect to the individual scores. Likewise, this concept is used to rank predators and casualties in a communication network.

Assumption: One bullying message is considered for each user.

Predator: Person who has posted at least one bullying message.

Victim: User who has received at least one bullying message.

Objective: To identify and to rank the most dynamic user as Predator and Victim.

Presently, a ranking method using the HITS module is utilized to detect predators and casualties. A user may be a predator and a victim depends upon on the

Figure 11.
Bullying network.

harassing messages he/she sends or receives. So, a user appointed as a predator and in addition with a casualty score. Predator and victim scores can be calculated by the following two equations.

$$p(u) \leftarrow \sum_{u \to y} v(y) \tag{13}$$

$$v(u) \leftarrow \sum_{y \to u} p(y) \tag{14}$$

Here, p(u) and v(u) are represented as the Predator and Victim scores respectively. u $\to y$ represents the existing harassing post from u to y, whereas $y \to u$ shows the presence of the bullying posting from y to u. The above equations are used for evaluating predator and casualty scores and also considered as repeatedly upgrade a set of equations. They depend upon the presumption that the most dynamic predator connects to the most dynamic victims by sending harassing posts. The most active victim is connected to the most dynamic predators by getting bullying messages. Basically, the user's predator score increases when the user (u) is connected with another user with a high victim score. In the same manner, the user's victim score increments when the user (u) is connected through received bullying messages to a user with a high predator score. The scores are computed through incoming degrees and outgoing degrees, and associated scores, in each and every iteration and this may give the result in large values. Subsequently, scores are standardized to unit length, i.e., each predator and victim scores is divided by the sum of all predator and victim scores respectively.

Then there is a necessity to define the ranking methods to the predators and victims which is depicted in the network diagram in **Figure 11**. In order to explain a real scenario in a simple manner, only five users are selected as depicted in **Figure 12** as an example and it depicts the recognition of the most dynamic predators and casualties in a bullying network. It is a weighted directed graph G = (U,A) with a set of nodes are represented as |U| and a set of arcs are represented as |A| where,

Each node $u_i \in U$ is a user involved in the bullying conversation,

Each arc $(u_i,u_j) \in A$, is defined as a bullying message sent from u_i to u_j,

The weight of arc (u_i,u_j), denoted as w_{ij}, is defined as a summation of in-degrees. Predators and victims are recognized by the directed graph G with weight. The victim can be recognized with many incoming arcs and the predator can be recognized with many outgoing arcs of the respective nodes. This method is helpful to observe the most dynamic predator or a casualty.

3.6.2 Cyber bullying matrix

A cyber bullying matrix(w) is constructed to discover a predator and victim depends upon their individual scores. It is depicted in **Table 7**. It is formulated as a square adjacency matrix (it represents the incoming degrees and outgoing degrees of each node) of the subnet with entry w, which is a square adjacency grid of the sub collection with entry w_{ij}, where,

$$w_{ij} = \{n \text{ if there be n harassing posts from } u_i \text{ to } u_j, 0 \text{ otherwise}\} \quad (15)$$

Since each client will have a casualty as well as a predator score, scores are represented as the vectors of n*1 dimension where i^{th} coordinate of the vector represent both the scores of the ith user, say p_i and v_i respectively. To calculate scores, equations p(u) and v(u) are shortened as the casualty and predator renovating matrix–vector multiplication equations. For the preliminary iteration, p_i and v_i are started at 1. For every client (say, i = 1 to N) predator and victim notches are as follows:

$$p(ui) = wi1v1 + wi2v2 + \ldots + wiNvN \quad (16)$$

$$v(ui) = wi1p1 + wi2p2 + \ldots + wiNpN \quad (17)$$

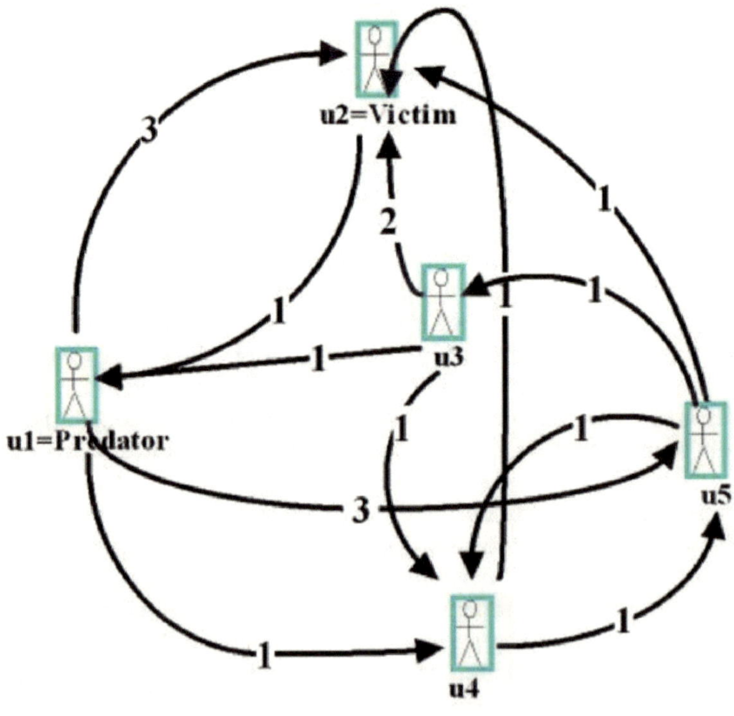

Figure 12.
Communication paths between predator and casualty.

Sender	**Recipient**						
	U_1	U_2	U_3	U_4	U_5	… …	U_N
U_1	0	3	0	1	3	….	….
U_2	1	0	0	0	0	….	….
U_3	1	2	0	1	0	….	….
U_4	0	1	0	0	1	….	….
U_5	0	1	1	1	0	….	….
… ..	….	….	….	….	….	….	….
U_N	….	….	….	….	….	….	….

Table 7.
Cyber bullying matrix (W).

When these equations congregate at a stable value (say k), it offers the final predator and casualty vector of each user. At last, to compute the eigenvector to acquire the predator and casualty scores.

Algorithm 1 gives a general framework of identification of the top-ranked most active predators and victims. In the algorithm N is a total number of users and Top is a threshold value, which is set manually.

Algorithm 1. Predators and casualty recognition.

Input: Set of consumer engaged in the chat with harassing post, N, Top.
Output: Set of Top Casualty and Top Predator

1. Take out dispatchers and receivers from N;
2. Initialize predator and casualty vector each N;
3. Generate adjacent matrix w using formula (15);
4. Compute Predator and casualty vectors with iterative updating Eqs. (16) and (17), and normalize, until congregate at secure value k;
5. Compute Eigen vectors to locate Predator and Casualty scores;
6. Revisit high ranked Predators and Casualties.

4. Summary

The new system is achieved by two commitments. First, a Novel Statistical Application, which is established on the new Bully-LDA with the weighted B-TFIDF strategy on bullying like attributes. It also efficiently and effectively finds latent bullying features to cultivate the accomplishment of the classifier and also to reduce the feature sparsity. Secondly, a Graph Model lends a hand to pinpoint the attackers and causalities in social networks. Such a system would encompass the following function: Tweets Crawling, Tweet Preprocessing and Tokenization, Feature extraction and Frequency extraction, Text Representation Model, Text Classification, Category of Texts, Performance Evaluation, and Results.

The Twitter corpus consists of text communications by way of metadata such user ID, dispatching time, etc. Tweets Crawling is performed using many classes and techniques in order to get the information of the users' connected data and the details of the Tweets' which is done using Twitter's Application programming interface called "Twitter4j-core-4.02.jar." Tweets are shown in entirely colloquial manner, with more amount noise and variation in linguistics. For example, tweets contain a hefty quantity of novel words, interjections, repetitions, short words such as acronyms, words with missing letters, words with phonetic spelling like *Gud* for *Good,* etc. and also missing blank spaces between the words, such as *whatareyoudoing,* which increases the tweet length. All these things impose a huge burden in the analysis of the text. Text preprocessing module contains word segmentation, word processing, and subsequent analytical steps include like converting uppercase letters to lower case, stemming, eradicating stop words, superfluous characters and hyperlinks.

The proposed framework utilizing Bully-Latent Dirichlet Allocation through Support Vector Machine has been examined with Twitter messages. This system is based on a novel concept of applying text mining techniques to tweets for detecting Bullying messages and also to identify Predators. The weighted B-TFIDF function is used to enhance the execution of classification, in which bullying-like features are measured. The overall results using Bully-LDA + SVM and weighted B-TFIDF outperformed other models. This model has numerous benefits adding more accuracy, superior noise diminution, faster speed and greater automation. The results obtained were analyzed properly using different metrics. A range of performance measures for instance accuracy, recall and F1 measures were calculated. The analysis of results plainly displays that the system yields effective results in identifying bullying messages in a successful manner.

In this research, a methodology for cyber bullying recognition of the most operative predators and casualties are done powerfully and fruitfully. This chapter presents a framework for detecting cyber bullying in Twitter using Bully-Latent Dirichlet Allocation with support vector machine. The preprocessing procedures have pertained to tweets. First Bully-LDA, a statistical topic modeling is used on a massive Twitter Corpus, with the help of weighted B-TFIDF scheme to detect offensive words in tweets. Next, a graph representation is utilized to recognize the predators and casualties in Twitter.

Author details

K. Nalini[1]* and L. Jabasheela[2]

1 Bharathiyar University, India

2 Panimalar Engineering College, India

*Address all correspondence to: immanuelsamen@rediffmail.com

References

[1] Ibn Rafiq R, Hosseinmardi H, Han R, Mishra S, Lv Q. Scalable and Timely Detection of Cyber Bullying in Online Social Networks. France: ACM; 2018. pp. 1738-1747

[2] Zhong A, Li H, Squicciarini A, Rajtmajer S, Griffin C, Miller D, et al. Content-driven detection of cyberbullying on the Instagram social network. In: Proceedings of the Twenty-Fifth International Journal Conference on Artificial Intelligence, (IJCAI-16). 2016

[3] Sherly TT, Rosiline JB. Supervised feature selection based extreme learning machine (SFS-ELM) classifier for cyberbullying detection in Twitter. International Journal of Scientific and Research Publications. 2017;**7**(7): 367-373

[4] Gotardo MA. Topic modelling of online child pornography documents. International Journal of Social Science and Economic Research. 2018;**3**(2): 505-521

[5] Priyangika S, Jayalal S. Detection of cyberbullying on social media networks. In: International Research Symposium on Pure and Applied Sciences. 2016

[6] Kansara KB, Shekokar NM. A framework for cyberbullying detection in social network. International Journal of Current Engineering and Technology. 2015;**5**(1):494-498

[7] Parapar J, Losada DE, Barreiro Á. Combining psycho-linguistic, content based and chat-based features to detect predation in chat rooms. Journal of Universal Computer Science. 2014; **20**(2):213-239

[8] Huang Q, Singh VK, Atrey PK. Cyberbullying detection using social and textual analysis. In: Proceedings of the 3rd International Workshop on Socially-Aware Multimedia, ACM. 2014. pp. 3-6

[9] Zhao R, Zhou A, Mao K. Automatic detection of cyberbullying on social networks based on bullying features. In: Proceedings of the 17th International Conference on Distributed Computing and Networking, ACM. 2016. p. 43

[10] Chen Y, Zhou Y, Zhu S, Xu H. Detecting offensive language in social media to protect adolescent online safety. In: Privacy, Security, Risk and Trust (PASSAT), International Conference on Social Computing (Social Com), IEEE. 2012. pp. 71-80

[11] Steyvers M, Smyth P, Rosen-Zvi M, Griffiths T. Probabilistic author topic models for information discovery. In: Proceedings of the 10th ACM International Conference on Knowledge Discovery and Data Mining, ACM Press, New York. 2004. pp. 306-315

[12] Blei D, Gri T, Jordan M, Tenenbaum J. Hierarchical topic models and the nested Chinese restaurant process. In: Seventeenth Annual Conference on Neural Information Processing Systems, NIPS. 2003

[13] Griffiths TL, Steyvers M. Finding scientific topics. Proceedings of the National Academy of Sciences of the United States of America. 2004;**101**(1): 5228-5235

[14] Minka T, Lafferty J. Expectation-propagation for the generative aspect model. In: Proceedings of the Eighteenth Conference on Uncertainty in Artificial Intelligence. 2002. pp. 352-359

[15] Bastian M, Heymann S, Jacomy M. Gephi: An open source software for exploring and manipulating networks. In: International AAAI Conference on Weblogs and Social Media. 2009. pp. 361-362